THE BALD EAGLE

Haunts and Habits of a Wilderness Monarch

Jon M. Gerrard and Gary R. Bortolotti

Smithsonian Institution Press
Washington and London

*To all those—family and friends—who have given generously of their
time and effort in the study of eagles in Saskatchewan, for without them
there would not have been a Besnard Lake project.*

Copyright © 1988 by Smithsonian Institution
All rights reserved

Designer: Janice Wheeler
Editor: Lorraine Atherton

Library of Congress Cataloging-in-Publication Data

Gerrard, Jonathan M., 1947–
    The bald eagle.
    (Smithsonian nature book)
    Bibliography: p.
    Includes index.
    Summary: Describes the morphology, behavior, flight patterns, hunting, migra-
tion, nesting, development, and growth of bald eagles.
    1. Bald eagle. [1. Bald eagle. 2. Eagles] I. Bortolotti, Gary R. II. Title. III. Series.
QL696.F32G47 1988   598'.916   87-26530
ISBN 0-87474-450-4
ISBN 0-87474-451-2 (pbk.)
British Library Cataloging-in-Publication Data available

Cover: *A Bald Eagles' nest at Besnard Lake. While the male broods the young eaglets, the
female arrives with some moss. Photo by Gary Bortolotti.*

∞ The paper used in this publication meets the minimum requirements of the
American National Standard for Permanence of Paper for Printed Library Materials Z39.48–1984.

Manufactured in the United States of America

Contents

Foreword

Jon Gerrard, Doug Whitfield, and a pilot they had hired are the only persons with whom I have ever shot a river rapids in a pontoon airplane. We were in a remote area of northern Saskatchewan surveying eagles in 1969 using new aerial searching techniques that Jon and Doug had been developing. After flying for several hours, we decided to land on the water for a break. Unfortunately, the pilot did not use the best judgment, and we landed on a stretch of river that was not large enough to take off from again. After several aborted attempts at taking off, we shut down the engine and sat to ponder our fate.

Finally, after considering the alternatives and studying the maps, Jon determined that if we could make it through an area of rapids to another part of the river, we would have enough room to take off. So we all got out on the pontoons and took out paddles, carefully paddled and maneuvered the plane into position above the rapids, and then shot the rapids. It worked, and we were able to take off and complete the eagle survey.

We were a long ways from any other humans or help, including radio contact, and that was before the days of emergency location transmitters that aircraft are now required to carry. If we had not made it out then, we all might still be there today! But many years and a few other mishaps since then, we now have better techniques for working with eagle populations and much accumulated knowledge, thanks in large part to the efforts of Jon, Doug, and others working with them in Saskatchewan and Manitoba.

I and a few others in North America had been using airplanes to get to eagle nests before 1969, but it was mostly just to visit nests that we already knew about. Jon and Doug, however, were using airplanes to search for new nests and to develop techniques that could yield better estimates of how many nesting pairs of eagles actually existed.

Bald Eagles are large birds, with wingspans up to eight feet, and their nests are also large. The average eagle nest is a mass of sticks about six feet in diameter and three to four feet from top to bottom, with the largest on record being nine feet in diameter and twenty feet from top to bottom. But the birds and their nests are spread over large expanses of forest and usually are difficult to see against the background of trees, so it is not easy to obtain estimates of their numbers. Jon and Doug were using new aerial techniques to find Bald Eagles and their nests. They had invited me along on their work to see and learn what they were doing. It was an eye-opener for me and greatly aided my own subsequent work. Their studies were similarly significant for others who have worked with eagles.

Following their earlier extensive surveys in Saskatchewan and Manitoba, Jon and his colleagues focused their research more intensively on a population of Bald Eagles at Besnard Lake in Saskatchewan. There they conducted a series of studies and involved other researchers in the work. One of them was Gary Bortolotti, originally a student at the University of Toronto. Gary proved his skills both in the field — climbing trees up to eagle nests, handling boats — and in the academic realm, designing studies, analyzing data, and publishing the results. Just as Jon and Doug's work had been significant, Gary's definitely advanced our knowledge and understanding of Bald Eagles. Gary's work also provided new research and measurement methods that I and other eagle researchers now routinely use.

Bald Eagles throughout much of North America were suffering severe reproductive problems during the fifties through the mid-seventies, largely because of chemical contamination from DDT. We now know, in part because of the contributions of Jon and Gary, that the species is making a strong comeback, and the future, while still not certain, is definitely brighter.

Thus, both Jon Gerrard and Gary Bortolotti are eminently qualified to write this book on Bald Eagles. They have woven a fascinating account that merges many of their experiences with the ecological and other biological information about the birds to which they themselves have contributed so much.

James W. Grier
Professor of Zoology
North Dakota State University

Preface

Raptors — falcons, hawks, eagles, and owls — fascinated me from the time I was quite young. A train journey across the Canadian prairies from Yorkton to Saskatoon when I was seven is still vivid in my memory. We saw many hawks on the way. One in particular, a Red-tailed Hawk, swooped so low and close beside the moving train that I could see the individual feathers of its brown back and red tail. The picture is still so clear it might have happened yesterday. In my imagination the hawk was coming down to catch a rabbit. In fact the moving train and the tall grass prevented my knowing the outcome of its flight. It could just as easily have been descending to land, or it might possibly even have been flushed by the train.

In the years after the train ride, I was fortunate to be invited on a number of bird banding trips by Dr. Stuart Houston, Saskatchewan's physician-ornithologist. One time, distraught at being left behind on the annual trip to band pelicans and gulls at Redberry Lake (I had twisted my ankle the day before), I protested so loudly that I was promised a trip with Dr. Houston to band Golden Eagles. A week later we drove south from Saskatoon to Lucky Lake to be guided by the unforgettable, half-blind, and aging Dave Santy. We went through cattle gates and over hilly rangeland to a coulee leading back from the South Saskatchewan River. The nest was fifteen feet below the cliff top on the side of the coulee and accessible from above. After the young had been banded by an experienced climber, I was lowered by rope into the huge nest to come face to face with a big black bird. It seemed at least half as big as I was. Poised on the side of the sand cliff in a nest full of rabbit bones, the young eagle — its talons spread wide — gazed at me, just as I studied it. All too soon a call from above signaled the end of my turn on the nest.

In May of 1966 on an expedition to band Great Horned Owls with Stuart Houston, we were joined by Doug Whitfield. Doug, from southern Ontario, grew up a modern-day voyageur. Already he had canoed several of the major Canadian fur-trade routes. Soon we were regular canoeing companions. Our attention gradually switched from the raptors of the prairies to those of the lakes and rivers of the northern forests — and in particular to Bald Eagles.

Early that fall we canoed in northwestern Ontario. Day after day we saw soaring eagles. Once, rounding a bend, we canoed almost beside a swimming eagle and watched wide-eyed as it clambered ashore in front of us. We left our canoe to walk into the tall grass to inspect it. The bird was thoroughly soaked. It looked frightened and quizzical as it gazed at us. It could not fly, although it did not appear injured. After a bit it ambled off into the bush. We left it, uncertain whether it would survive. Since then, however, we have on other occasions seen swimming eagles get so wet that they could not fly.

The idea of banding young eagles in their nests grew from a vague notion to a burning desire. The next summer, four trips to northern Saskatchewan launched our careers as eagleologists (at the time, I was in medical school and Doug a graduate student in physics). Funding from the Canadian Wildlife Service to fly surveys the following two years gave us a real boost. In subsequent years, we established first a tent camp, then a cabin-based project on Besnard Lake.

Though Doug has now retired as a climber of eagle nests, I have continued. One of the rewards of my work on Besnard has been my association with Gary Bortolotti, who first came to the lake in 1976 and then returned from 1979 to 1982 to do doctoral research. Gary, a fiery, enthusiastic Torontonian, has joined me in writing this book.

For the nine years from 1971 to 1980, I lived in winter in Minneapolis, Minnesota. During that time I was fortunate to visit many of the sites where wintering eagles congregate. Seeing the birds in all seasons helped immeasurably when it came to writing this book.

Each chapter begins with a personal description of our experiences. We felt that in this way we could convey some of the excitement of seeing Bald Eagles in their natural habitat. At the same time, each individual description gives insight into eagle behaviors that are described more fully in the remainder of the chapter. Though we have drawn upon the experience and hard work of other biologists, this book—a look at the life of these birds and a tour through the eagle country of North America—emphasizes our personal encounters. Since the descriptions at the beginning of each chapter are personal, we have noted at the start of each one whether it is contributed by Gary or myself. The first draft of the remainder of each chapter was written by the same author, but in some cases revision and rewriting have been such that the final version can only be considered an amalgam of our efforts.

J.G.

Acknowledgments

Our close friend and colleague Doug Whitfield and Jon's colleague-wife, Naomi, are first and foremost on our list of people whose contribution to our eagle research must be recognized. From early on they have been equal partners in the eagle project. During the last twenty years, many others have worked with us in the field or provided critical help in other ways: Vernon Arnold, Don Buckle, John and Marilyn Curry, Joe and Sharon Daly, Wayne Davis, Elston and Connie Dzus, David Eagen, Nancy Flood, Maggie Gazziano, Jim and Perry George, Peter, Nikki, and Chris Gerrard, Robin Goodfellow, Henry Halkett, Al Harmata, Ginny Honeyman, Sara Israels, Mike Kipley, Carol Kline, Graham Leathers, Ted and Anna Leighton, Bill Maher, Michelle Mathys, Arnold Moen Nijssen, Alan Moulin, Chet and Miriam Meyers, Jim and Judy Potter, Audrey Remedios, Rick Sanderson, Jeff Stafford, Jennifer Staniforth, Jack Stilborn, Rick Stone, Kandyd Szuba, John Toepfer, and Tom Wilson. We thank John and Demar Hastings, the Sims, and the Hilliards of Besnard Lake for all their neighborly help and company. Jon's parents, John and Betty Gerrard, deserve special mention for their continued support. Many other people gave generously of their time and hospitality in our pursuit of eagles across North America.

Research on eagles can be a volunteer effort only so long before costs become prohibitive. Initially our work was funded by the Canadian Wildlife Service. Since then we have received financial support from the World Wildlife Fund (Canada), the Eagle Valley Environmentalists (now the Eagle Foundation), the National Wildlife Federation, the Hawk Mountain Sanctuary Association, and the Natural Sciences and Engineering Research Council of Canada in a grant to Jon Barlow. Mrs. H. E. Henderson of Montreal deserves special thanks for her private contributions. Logistical support in the form of equipment was provided by the Department of Northern Saskatchewan, Nikon Canada, and the Outboard Marine Corporation.

Gary is especially indebted to his academic advisor and friend Jon Barlow, of the Royal Ontario Museum and University of Toronto, for providing him with the opportunity to work on eagles for a Ph.D.

For their comments on various chapters we thank Viola Dufresne, Elston Dzus, Elizabeth Gerrard, Naomi Gerrard, Linda Hutcheon, Kandyd Szuba, and Heather Trueman. Keith Cline and Maurice LeFranc were helpful in providing access to the files of the National Wildlife Federation. The lines of poetry at the start of Chapter 1 are reprinted by permission, © 1952 *Saturday Review* magazine, and the lines at the start of Chapter 12 are reprinted by permission, © 1966 The New Yorker Magazine, Inc.

Teryl G. Grubb, Mark McCollough, Peter E. Nye, John E. Swedberg, Clayton White, and Frank Wille kindly provided photographs. Naomi Gerrard provided us with artwork. Natalia Eyolfson assisted with the typing.

Our immediate families — the Gerrards; Naomi, Pauline, Charles, Thomas; and the Bortolottis: Heather and Lauren — are recognized for their unending, selfless support through all phases of our lives.

THE BALD EAGLE

Haunts and Habits of a Wilderness Monarch

An adult Bald Eagle perches beside its nest at Besnard Lake, Saskatchewan (Gary R. Bortolotti).

1. Historical Reflections

Solitary by the sun,
The bird that has been empires
Knows his time has well-nigh run.
Robert P. Tristram Coffin

Jon —

Splat! Our windshield darkened momentarily as the Volkswagen Beetle plunged into mud. Almost before the thick, brown ooze covered the glass, Doug had the wiper in action. As the car slid into the mire, it stayed on course for a moment. The front end swung slowly to the right. Doug jerked the steering wheel left. The car swung back. For a fleeting second it faced straight ahead, then continued on pointing progressively farther left. Doug, anticipating this, was already turning the steering wheel furiously clockwise. Oscillating back and forth, we traversed the mudhole and slowly climbed the slippery grade on the other side. At the top of the rise the car was finally stabilized, and almost stopped. We looked ahead. Half a mile in front was another muddy spot. Doug sped up — we needed momentum to carry us through.

For two hours we battled the mud of this central Saskatchewan road. Finally, we arrived at Lake Mistiwasis. We were looking for, among other things, eagles. Just before we reached the lake, another road, in much better condition, came in from the east. The lake itself had cottages all around. There were no eagles. We would have to seek a more remote lake farther north.

A few weeks later and two hundred miles north, at Otter Lake on the Churchill River, we removed the canoe from the roof rack and were soon paddling across windblown waters. At the other side of this wide outpouching of the Missinipe River, the "Big River," as the Churchill is called by the local Cree, we portaged into a series of small lakes. Toward evening we camped. The next day, as we paddled along a narrow stream, we saw, fleetingly and far away over a ridge, a dark brown eagle. Four days later, saturated in the sun, the wind, and the clear blue water of the lakes but disappointed to see no more eagles, we returned to our starting point.

The following year, 1967, armed with better information on eagle nest sites obtained from various people who had visited the northern parts of the province, we returned to the country of the Missinipe. As we paddled our canoe we imagined we were voyageurs traveling into little-known regions. This time, on Dead Lake, we found what we were looking for: a huge mass of sticks high in a poplar tree. Two adult eagles perched nearby. Two half-grown young peeked over the rim. Doug strapped spurs on his feet, climbed the tree, and put Wildlife Service bands on the right leg of each of the six-week-old youngsters. The next day, at Nemeiben Lake, it was my turn to climb. From there, we continued—Candle Lake, Squaw Rapids, Hansen Lake, Pelican Lake. Each new nest was a new discovery. Our exhilaration grew. Canoeing, however, was time-consuming. Twice we rented fishing boats; although the romance of mechanized travel was less, the twenty-horsepower motor took us much more swiftly to our objective. The last weekend, through a stroke of luck, we received a ride in a float plane courtesy of the provincial Department of Natural Resources. At each of 4 nests, our pilot brought the Cessna down on the nearby lake. With the plane tied up beside a flat open rock or on a sand beach, we walked the short distance into the forest and climbed to the nest. This was the way to band eaglets in style. The score that year was 27 young found in 18 nests. Every tree had been climbed; every young had been banded.

The next year, with funding from the Canadian Wildlife Service, we flew much more. In a small plane weaving back and forth along the shoreline just above the treetops, we traveled in the air with the eagles. The myriad lakes, as numerous as the leaves on a tree, stretched almost endlessly in all directions. By the end of three weeks we had charted the locations of 142 nests. Besnard Lake, with 8 nests found in a quick search, was particularly productive. Here, in the years ahead, we would thrill again and again to see, to study, and to understand these majestic birds.

We will never know with any certainty how many Bald Eagles were present in North America when Europeans first arrived. The eagles nested on both coasts and along every major river and large lake in the interior from Florida to Baja California in the south and from Labrador to Alaska in the north.[1] In many areas, they were abundant. F. Kirkwood believed that there was a Bald Eagle nest for every mile of Chesapeake Bay shoreline as late as 1890 (though likely not all those nests were occupied by breeding pairs). A similar nesting density may have existed along the shores of the Great Lakes. Even away from the Great Lakes in Michigan "probably a few pairs nested in every county." The situation was similar in Minnesota, where one or more pairs nested on nearly every large

Bald Eagle Breeding

When Europeans arrived in
North America

In the late 1960s

*Breeding range of the Bald Eagle as it was in the late sixties (black area), and as it was
when Europeans first arrived (black plus stippled areas).*

lake in the state. In the south, in Louisiana, Bald Eagles were "generally
distributed" in the cypress swamps of the southern part of the state. Farther west, in Colorado, they were "nesting abundantly in the mountain
parks"; and on the west coast, it was "one of the most abundant birds of
the Falcon Tribe in Washington Territory." In the early 1800s, John
Richardson noted that in Canada, from Lake Superior to Great Slave Lake,
the Bald Eagle "abounds in the watery districts of Rupert's Land."

Observations in winter support the evidence from breeding areas.[2] On
Manhattan Island, New York, in the mid-1800s the Bald Eagle was
"extremely abundant on the floating ice of the [Hudson] river and sometimes brought its captive fish to the trees in the park, there to eat them or
as often to quarrel about them with its fellow." At about the same time
along the Mississippi River, near Keokuk, Iowa, the air was "simply alive"
with eagles feeding on offal discarded by the pork houses. We suspect
there were between a quarter million and a half million Bald Eagles on the
continent when Europeans first arrived. What has happened since?

The first major decline in eagle numbers probably began in the mid- to late 1800s, starting in the east and progressing westward, coinciding with the movements of white settlers.[3] Audubon, one of the first to note a decrease, commented that the White-headed Eagle, once "extremely abundant" in the lower parts of the Ohio and Mississippi rivers, had been much diminished by the 1840s. He believed that man's persecution of the game on which the eagles fed was to blame. On the western plains, from Montana to Oklahoma, eagles were disappearing by the late 1800s. Observant naturalists living in central Montana from 1898 to 1903 and along the lower Yellowstone River in the southern part of the state from 1888 to 1907 failed to report a single Bald Eagle.

Part of the reason for the decrease in Bald Eagle populations late in that century may lie in the complete or virtual extermination of species like the buffalo and the Passenger Pigeon. Although there is no proof of a link between eagle and bison or pigeon populations, we speculate that a major food source for eagles disappeared with the loss of the great herds and flocks. Today, carrion is important to the survival of eagles in winter (see Chapter 11). Winter food must have been abundant when the huge bison herds numbered in the millions. Similarly, immense flocks of Passenger Pigeons and incredible numbers of ducks, geese, shorebirds, and seabirds present in the early 1800s must have provided food for eagles.[4]

As settlement moved west across the continent, farms and ranches began to appear on what had been virgin territory. The countryside swarmed with people with guns. A threat to livestock or the collection of a trophy was an excuse for some marksmen; an easy target was good enough for others. In 1888 Barton Evermann in Illinois commented, "Scarcely does an eagle come into our State now and get away alive, if he tarry more than a day or two."[5] In Montana, E. S. Cameron noted that the diminishing population of eagles was all but eliminated by 1905 as a result of high bounties placed on wolves. Traps set for wolves sometimes caught eagles. Often though, a "wolfer" would travel from one creek to another. "Deer and antelope were shot down wholesale in a line across the country, the carcasses filled with strychnine, and the poisoned bait scattered."[6] From one end of Montana to the other, eagles, as well as wolves, were lured by an easy meal and died convulsive deaths, their talons spread apart and their bodies contorted.

In some areas, the native people were suspected of depleting eagle numbers.[7] The Bald Eagle was apparently quite common in the Wichita Mountains of Oklahoma but was almost extinct by the time J. H. Gaut visited there in 1904. He suggested that persistent killing by Indians for the birds' feathers, prized for ceremonial bonnets, was the cause of the decline. Indians on occasion also killed eagles for food, though probably only when other game was not available.

By the early 1900s, from the Ohio River in the east all the way west to the Cascade Mountains in Washington State, the Bald Eagle was rare where it had once been common or abundant.[8] Even in Florida, which managed to keep a considerable number of eagles, Oscar Baynard noted in 1913 that the species was "not holding its own now as every hog raiser in the county [Alchua] kills every one he can on account of the Eagle's perverted taste for razor back pig."[9] In Canada, except for the seacoasts, the Bald Eagle had become "nothing more than a rare, interesting, and picturesque feature of the landscape."[10]

The end of the Bald Eagle as a common feature of the American landscape was characterized by Henry W. Shoemaker in 1919:

> Suspended by a dirty string
> In a dingy down-town store
> With wings wide-spread
> A stuffed Bald Eagle hangs
> And as the summer breezes blow
> Filth-laden through a small window
> The regal bird, which once did soar
> Above the clouds, above the storm
> Swings gently round and round.[11]

Direct slaughter, removal of important winter foods, and changes in habitat, especially a decrease in suitable nesting habitat, appear to have kept Bald Eagle populations low until the forties.[12] In the thirties and forties, records for several states indicate, eagle populations were beginning to recover.[13] A combination of factors was likely responsible. The Bald Eagle Act was passed in 1940 to provide protection in the United States (excluding Alaska, where there was a bounty on Bald Eagles because they were suspected of competing with fishermen for salmon). A new public awareness started and direct persecution lessened. The construction of locks, dams, and reservoirs along the Mississippi, Missouri, and other rivers began in the twenties and thirties, and the creation of wildlife refuges provided improved winter habitat. Here, fish and waterfowl, upon which eagles feed, congregate and are perhaps more vulnerable. Musselman, watching eagles wintering along the Mississippi River below the dam near Keokuk, Iowa, was the first to note the recovery. In the thirties he usually saw 5 to 10 eagles but wrote of many that had been shot. In 1940, for the first year in a decade, no eagles were reported shot. By 1945 there were 25 to 30 eagles wintering regularly in the region; in 1948 it was up to 59 birds. By 1979 eagles had increased to the point that more than 400 were counted at one time.

Observations during late fall in Glacier National Park, Montana, suggest that recovery there started a few years later. On average, 22 birds were seen each year from 1937 to 1950.[14] An increase was evident by 1951, with 59 eagles seen that fall. By 1954 the total was 86, and by 1978 it had surpassed 600. A consistent pattern of increased numbers was evident in reports from widely separated areas by the fifties. Across interior western North America, eagle populations were recovering. At the time of the ini-

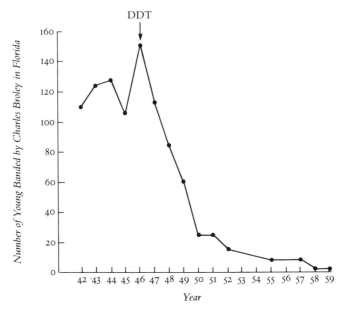

The number of nestling Florida Bald Eagles banded by Charles Broley began to decline dramatically after the introduction of the pesticide DDT in 1946. The decline reflects a decrease in the productivity of eagles, not in Broley's effort.

tial decrease in Bald Eagle numbers in the region, both the local breeding population and the migratory population from Canada were affected. The size of the migratory population was probably historically larger than the local breeding population; breeding in more remote areas, the migrating birds contributed most to the rebound.[15]

However, not all the news about eagles in the western states was good. In the sheep-ranching country of Texas, ranchers hired gunners to shoot eagles from airplanes. From 1950 to the mid-sixties, 20,000 Golden and Bald eagles probably died in this massacre. Other eagles were poisoned in Wyoming. In spite of that, the upward trend in the number of Bald Eagles wintering in the west continued through the sixties and into the seventies as improved enforcement of the Bald Eagle Act began to curtail the wanton slaughter. Quite good evidence suggests that the eagle population wintering in the interior western states doubled between 1955 and 1980.[15]

Alaska was exempted from the Bald Eagle Act. There, by the early fifties, 128,000 pairs of eagle legs had been turned in for bounties. How many more birds were shot but not recovered we shall never know. Finally, after lengthy studies concluded that Bald Eagles were not having an adverse effect on salmon numbers, in 1952 protection reached Alaska. By 1980, if not before, eagles had recovered to the point of saturating the Alaska habitat.[16]

Just as there were signs that some eagle populations were increasing, a new threat emerged. Charles Broley, a retired banker from Winnipeg, Canada, was the first to notice.[17] In 1939 he had begun banding eagles along the Gulf coast of Florida from Hernando County south to Fort Myers. By 1946, he knew of 140 active nests. In 105 of them, he banded 150 eaglets. In 1947, eagle productivity was low compared with previous years. That was just the beginning. So drastic was the decline that by 1952 Broley was able to band only 15 eaglets in 11 nests in the same locale. Many adults were still present, but they failed to produce young.

Broley began to suspect that pesticides were to blame for the decline in eagle productivity. He was right. Shortly after the Second World War, DDT had been sprayed extensively along the coast to control salt marsh mosquitos.[18] DDT accumulated in the prey of Bald Eagles and other predators to such levels that the birds could no longer reproduce. Some eggshells were too thin to support the weight of the incubating parent. Other eggs were infertile or contained dead or deformed embryos. The effect of DDT soon extended far beyond Florida. To Broley's dismay, when he tried to band eagles in Ontario near his summer cottage in 1951, reproductive failure was widespread. Charles Broley died in 1959, but Rachel Carson and others took up and pursued the battle against DDT.[18] Industry and government were slow to respond. It was not until 1970 that the use of DDT was curtailed in Canada, and it was banned in the United States on December 31, 1972. For some birds it was too late. The Peregrine Falcon was extinct as a breeder in eastern North America except for the Arctic. East of the Mississippi River the Bald Eagle barely held on, surviving (except for isolated pairs) as a breeding bird only in parts of Florida, the Chesapeake Bay region, Maine, and interior Michigan and Wisconsin. Populations breeding in the western United States were also affected, though those breeding in Canada from Saskatchewan west and in Alaska were spared.

Today the more pristine populations, or those more fully recovered from historical decimation, are doing well in Alaska, in coastal British Columbia, and in the boreal forest and the maritime provinces of Canada. In British Columbia and Alaska, areas of extremely high density, there may be as many as 50,000 eagles. Saskatchewan may have 12,000.[19] The remnant clusters of breeding eagles that survived the DDT debacle are now making a remarkable comeback. For example, the total production in the Chesapeake Bay region has increased from a low of 7 young in 1962 to 188 young in 1986.[20] Programs to restore populations by releasing young eagles (see Chapter 12) and to protect critical habitat are currently helping to extend the improvement throughout North America.

2. To Know an Eagle

But you never were made, as I,
On the wings of the winds to fly!
The eagle said.
Will Carleton

Gary—

I winced at the clang; yet another rock bounced off the oil pan of Doug's Volvo. Our bodies strained against the seat belts as we bumped and rattled our way down the gravel logging road that led us to our destination—Besnard Lake. During this last leg of our journey from Toronto, the vast, golden prairie gave way to open aspen parklands, which in turn gave way to the densely treed boreal forest. Here, we passed seemingly endless miles of pure stands of jack pines—all the same height, all the same diameter. They were replaced by seemingly endless miles of black spruce trees—all the same height, all the same diameter. The pines returned and, in time, so did the spruce as the forests alternated across the landscape. Although it is almost disconcertingly monotonous to many people, I somehow found comfort in the simplicity of the boreal forest. Perhaps it was an illusion of a landscape untouched by the ravages of humankind. My mind was free from the distractions of the hard-sell billboards and pit stops of the south. The monotony of the boreal forest was a verdant backdrop to my accelerating imagination as we neared the lake. Today, at long last, I would see my first Bald Eagle.

My emotions were strong and strangely polarized. On the one hand I saw myself as an intruder, an unwelcome tourist in a foreign land. This wild and beautiful country was so different from the metropolis of Toronto where I grew up that it might as well have been on the other side of the globe. On the other hand I had never felt such a sense of satisfaction, or perhaps relief, the way one feels when finally starting on the journey home after an extended absence.

Just three years ago I had received letters from Rick, a forestry classmate, whose summer job had been to construct the very road on which I was traveling. Rick had written in wonderfully sensitive prose of the beauty of the northern wilderness. I remembered the agony of envy as I read the words "This lake even has a Bald Eagle's nest!" Long had I dreamt of seeing Bald Eagles. Never did I really believe that I would have the chance to see them nesting, let alone to study them. Now, here I was, about to step on the shore of a lake with more than a score of nests. I was shaking with excitement. A few short months ago I had learned of the work that Jon Gerrard and Doug Whitfield had been doing on eagles at Besnard. A long-shot letter to Jon with an offer to volunteer my services for the summer paid off. Jon and Doug scraped together the funds to feed me and a friend. Before I knew it, the lake was within reach. I was to learn many things at Besnard Lake, not the least of which was that dreams can come true.

*W*hat is this bird, the Bald Eagle, that we had come so far to study? The image of an eagle arouses as many different feelings as there are people; few birds have elicited such strongly opposing emotions as these predators. Worldwide, eagles have been revered as symbols of majesty, grace, and power, inspiring poets and politicians alike. Yet all species have suffered from senseless, often relentless persecution. It was no doubt the sense of power and grandeur associated with the Bald Eagle that in 1782 inspired the people of the United States of America to adopt this bird as their national symbol. It is certainly an impressive-looking bird. Whether perched motionless on a dead branch above a river, its head glistening pearly white in the sun, or soaring gracefully in a light wind along a cliff, its wings spreading dark and wide in the breeze, this bird is truly awe-inspiring.

Powerful yellow feet and sharp black talons are the principal tools a Bald Eagle uses to secure its prey. It can tear flesh from a carcass with the aid of its large, hooked bill. Concerns that these weapons might be used to kill prey in competition with human hunters, to raid a farm of its livestock, or even to carry off a human infant (never reliably documented) have historically led to the persecution of eagles. The tendency of Bald Eagles to rob the smaller and weaker Osprey and to feed on carrion added further weight to the argument by Benjamin Franklin and others that this was a bird of "bad moral character."[1]

Just 22 years before being adopted as America's national symbol, *Haliaeetus leucocephalus* (literally, "sea eagle with a white head") was first described to science by Carl von Linné (Linnaeus), the father of modern taxonomy, based on a specimen sent to him in Sweden. Some early naturalists at first considered young Bald Eagles in their all-brown, immature plumage to be a different species than the White-headed, or Bald (a synonym for white, not hairless), Eagle. Even the great naturalist and artist John James Audubon was confused. His is a curious story. In his monu-

*J. J. Aububon's painting of the Bird of Washington,
or Washington Sea Eagle (courtesy of the New-York
Historical Society).*

mental art folio of 1827, *The Birds of America*, Audubon painted a giant eagle he dubbed the Washington Sea Eagle, or Bird of Washington. He thought it was a species new to science. It looked very much like a young Bald Eagle — the feathers of the body and wing were shades of brown, the bill was bluish black, the cere (the fleshy part above the bill) was yellowish brown, and the feet were orange yellow — but its dimensions were enormous. Its wingspan of 10 feet 2 inches exceeded that of the Bald Eagle by at least 3 feet!

Audubon shared Benjamin Franklin's low opinion of Bald Eagles ("They exhibit a great degree of cowardice," and "Suffer me, kind reader, to say how much I grieve that it should have been selected as the Emblem of my Country").[2] His great Bird of Washington was a different matter. In referring to the species' namesake, George Washington — clearly one of Audubon's idols — he wrote:

He was brave, so is the Eagle; like it, too, he was the terror of his foes; and his fame, extending from pole to pole, resembles the majestic soarings of the mightiest of the feathered tribe. If America has reason to be proud of her Washington, so has she to be proud of her great Eagle.[2]

Audubon actually had few encounters with his Bird of Washington. He first caught sight of the "species" in 1814 while ascending the Upper Mississippi. After years of searching for this rare bird, he finally shot one. Unfortunately, the specimen has long since disappeared. We are left with the artist's rendition, a few measurements, and an emotion-filled account of the bird in the wild.

Today taxonomists believe that the famous naturalist probably described an immature Bald Eagle, writing down its measurements from memory and exaggerating them in the process.[3] Assuming that a giant eagle never existed and that Audubon had no ulterior motive, there is an explanation that can at least help us to understand the origin of his confusion. First, we must consider how Bald Eagles vary in size, shape, and color according to age, sex, and geographic origin.

The Washington Sea Eagle bears a striking resemblance to the juvenile (less than 1 year old) Bald Eagle in coloration. The similarity is not with older birds, for both the feathered and unfeathered parts of the body of the Bald Eagle change color as a bird ages. The juvenile's bill and eyes are both uniformly dark. Over the course of 3 or 4 years, the bill lightens to a swirl of light and dark brown, then a yellowish brown, and finally to the bright yellow of the mature bird. The eye, initially sepia in color, lightens to a buff yellow (by age 1½ years), light cream (age 2½), and pale yellow (age 3½).[4]

The juvenile Bald Eagle (less than one year old) has an all-brown plumage except for the underside of the wings (Gary R. Bortolotti).

Three half-year-old (first winter, known age) Bald Eagles at a wintering site in Maine (Mark A. McCollough).

The overall appearance of the body changes as feathers are gradually replaced over the course of each summer. When they leave the nest, young Bald Eagles are generally dark brown or have shades of medium to dark brown, with the exception of their underwing linings, which are primarily white. Plumage changes after that point are complicated and confusing. In the past, plumage classification schemes had to be based on suppositions or on observations of captive birds, which may not mature in the same way as wild ones. More recent knowledge comes from observing birds of known age (color-marked or banded) in the wild.[5] Data from wild birds suggest that the bright white feathers appear at a younger age than previously supposed. Generally, after the all-brown juvenile dress, the body takes on a blotchy appearance; there is often a dark "bib" on the upper breast with a white-speckled band below on the belly and a white triangle (pointing downward) on the back. Changes in the plumage of the head and tail are most consistent and most helpful in determining an eagle's age. The crown, initially dark brown, bleaches with prolonged exposure to the sun to a buff color during the first winter. The ear coverts, behind the eye, remain dark brown until the eagle is 2½ or 3½ years old, so as the crown and chin lighten progressively, the eagle appears to have a dark stripe through the eye. The crown and chin lighten to a tan color at 1 to 1½ years old, to smoke gray or buff white at age 2 to 2½ (together with the

A 1½-year-old Bald Eagle (known age) at a wintering site in Maine (Mark A. McCollough).

dark eye strip this creates the ospreylike plumage), to white with brown flecks at age 3 to 3½. By 3½ the tail is also white, but it often retains a dark terminal band. At age 4 to 4½, most Bald Eagles are almost indistinguishable from adults at a distance, but up close, brown spots are noticeable on the head and tail. (On rare occasions there are enough spots to make a bird look like a 3½ year old.)

It was commonly believed that the amount of brown spotting in the head and tail of adults (many birds have a few flecks) was an indication of relative age—the whiter the feathers, the older the bird. However, this seems unlikely. Al Harmata trapped and color-marked a Bald Eagle in adult plumage one winter in Colorado. Five years later (when it was at least nine years old) it was caught in a coyote trap as it migrated through southern Saskatchewan. When I examined the bird it still had several light brown streaks in the head and a few small dark brown blotches in the tail. But unlike the coloration of older immature eagles, these spots were only visible upon close inspection. The pattern of dark spots on the tail may be consistent over the life of an individual. We have found molted tail feathers under one nest on Besnard Lake that showed spots in the same location on a specific feather three years in a row.[6]

Color is not the only change that occurs with the first four or five plumages, nor is it the only reason why Audubon may have mistaken a young

Bald Eagle for a new species. Although eagles in their first year of life are skeletally as large as their parents, their tail and wing feathers are much longer. The one-year-old Bald Eagle exceeds its parents by 8 percent, 13 percent, and 23 percent in the length of primary, secondary, and tail feathers, respectively. (Appendix 1 gives some measurements of birds of different ages.) With each successive molt up to adulthood the flight feathers get shorter; therefore, young eagles are not only larger than adults, but they are also different in shape.[7] The long feathers also give immatures distinct aerodynamic properties. (See Chapter 3.) That may have further contributed to the creation of the Bird of Washington, for Audubon remarked that the flight of his giant eagle was very different from that of the White-headed (Bald) Eagle.

Other major sources of size variation in Bald Eagles, besides the changing lengths of feathers with age, may have contributed to Audubon's confusion. As is the rule for raptors (but more often the exception among birds in general), the female Bald Eagle is larger than the male. In northern Canada and Alaska, females usually weigh 10 to 14 pounds, whereas their mates are generally only 8 to 10 pounds. However, males are not just scaled-down versions of females. The depth (not length) of a female's bill and particularly the size of her feet and talons are proportionately much larger than a male's, compared with other parts of the body, such as the length of the flight feathers (Appendix 2). The difference in shape between the sexes is often much more noticeable than the absolute size difference.

Impressions of the size of birds in the field are often misleading, especially if there is a large degree of size variation within a species. For the Bald Eagle, such is the case because geography, as well as age and sex, influences size. The smallest Bald Eagles breed in Florida, and the largest in Alaska (probably in the Aleutian Islands). In between, there is a gradation of small to large from south to north. Geographic variation in size prompted taxonomists to identify two subspecies, or races: the Southern Bald Eagle and the Northern Bald Eagle. The boundary between them was arbitrarily set at 40 degrees north latitude. However, given the gradual nature of the size gradient and the lack of any other distinguishing feature, recognition of two races is questionable.

Eagles from widely separated breeding grounds may at times be found in the same area, for they can travel over great distances; northerns may fly to Florida, and Florida birds may visit Canada.[8] Aubudon no doubt saw a mixture of races, especially as he made many observations along the Ohio River, in the center of the size gradient. He could have easily believed in the existence of a gigantic eagle if he had seen an immature female Northern Bald Eagle beside the much smaller adult male Southern Bald Eagle. (Remember that binoculars were not standard bird-watching gear at that time.)

Although Audubon was mistaken about the species identity of the Bird of Washington, he was certainly correct about its genus, *Haliaeetus*, the sea

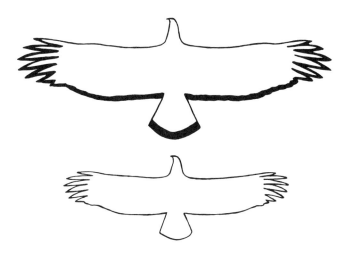

Comparison of the relative sizes of a female Northern Bald Eagle (top) and a male Southern Bald Eagle (bottom), drawn approximately to the same scale. The white area shows the size of an adult, and the black (on the Northern) shows the size of a juvenile, which has longer feathers.

eagles. Physical and behavioral traits link eight of the approximately sixty species of eagles around the world into this genus.[9] The sea eagles are typically large, powerful birds of seacoasts, lakes, and rivers. Most species subsist on a diet of fish, scavenging when they can, but they are also capable of taking waterfowl and mammals. Most sea eagles are social and may congregate in large numbers outside the breeding season. All but one species has a distinctive adult plumage with a striking pattern of white against brown and a juvenile plumage that is primarily brown.

Across Europe and Asia east to include Japan lives the White-tailed Eagle.[10] It is similar in size, appearance, and behavior to the Bald Eagle; in fact, the two are "superspecies," a term applied to species that are closely related but not so close as to be considered races of one species. The White-tailed Eagle has a more wedge-shaped tail and lacks the Bald's bright white head, but the juveniles of the two species are almost indistinguishable. Like the Bald Eagle, this species has suffered much from human persecution (it had been exterminated in Britain by 1916) and contamination of the food chain. The White-tailed Eagle also nests in western Greenland and Iceland and may sometimes be found (it has recently bred) on Attu, the extreme western island of the Aleutians. The only other sea eagle that occasionally frequents the North American shore is Steller's Sea-Eagle, the largest and most striking member of the genus. Its snowy white shoulder bands, tail, and thighs stand out against its dark brown body. Although at times the Steller's Sea-Eagle is spotted in the Aleutian, Pribilof, and Kodiak islands of Alaska, this rare bird usually haunts the Soviet coastline.

An adult (left) and juvenile (right) White-tailed Eagle from Greenland (Frank Wille).

One of the best-known *Haliaeetus* is the African Fish-Eagle.[11] The species' scientific name, *H. vocifer*, says something of this bird's habits. Its wild yelps fill the air when it defends its territory, and pairs even sing duets, throwing their heads up and down as they call. Bald Eagles, although also very vocal, are a bit more reserved; they do not duet, and only toss their heads straight up rather than throwing them all the way back to touch their backs, as fish eagles sometimes do. The Bald's repertoire largely consists of a call, or song, used in territorial encounters, a squeaky, disjointed alarm call, some grunts, and a loud wail used in aggressive situations (see Chapter 9).

The life histories of a few species—the Bald, White-tailed, and African Fish eagles—are relatively well known, but the remaining members of the genus—Sanford's Sea-Eagle of the Solomon Islands, Pallas' Sea-Eagle of Central Asia, the Madagascar Fish-Eagle of Madagascar, and the White-bellied Sea-Eagle of India, Southeast Asia, and Australia—are largely a mystery.

One feature of the sea eagles that makes them noticeably different in appearance from any other eagles is their massive bill. Developed to the extreme in Steller's Sea-Eagle, the deep, arched bill is unlike that of the Golden Eagle. There are several plausible explanations for this. Bald Eagles often feed communally, or at least compete with one another over large food items.[12] Other sea eagles likely do the same. Perhaps the large bill is a means by which large chunks of prey can be torn off and quickly swallowed. The Golden Eagle is a solitary hunter and so is not under the same pressure to feed in this manner. The bright coloring of the sea eagle's bill suggests that it is used as a signal in behavioral displays. For example, the Bald Eagle uses its huge yellow bill in courtship (see Chapter 8). Both large size and bright coloration would enhance conspicuousness for display purposes. More information on the behavior and ecology of other members of the genus is badly needed for comparison.

North America's only other resident eagle, the Golden Eagle, belongs to a group called the booted eagles, of the genus *Aquila*. Unlike sea eagles, their legs are feathered right to the toes. Although similar in size to the Bald Eagle, the Golden Eagle is ecologically, behaviorally, and evolutionarily distinct. The Golden Eagle is primarily a mountain and desert dweller, nesting on cliffs and hunting small mammals. Although sexually dimorphic in size, there is little or no difference in the size of flight feathers between young and old birds, and the difference in appearance between adult and juvenile plumages is much less marked than for Balds.[13]

Documentation of size, shape, and color variation, and of evolutionary affinities, is only the first step toward answering the question "What is a Bald Eagle?" It is an important step, however, for it sets the foundation for studies to come. We can now ask new questions about how age and sex might influence a Bald Eagle's behavior and ecology. Such research requires long, often frustrating and laborious, hours in the field. I for one could not spend my time better.

# 3. Flap, Glide, Soar

Eagles may seem to sleep wing-wide upon the air.
John Keats

Jon—

It has been calm all day. Cindy, an adult color-marked eagle, was perched silhouetted against the sky, atop a broken-off dead spruce. She looked like a sailor peering out from the crow's nest of a ship. The effect was enhanced by the presence of two other tall dead trees on the island. Together the three stood against the sky like the masts of a sailing vessel.

Cindy's perch tree must have been a giant at one time. Even with its top gone it towered above the neighboring live spruce on Kingfisher Island. Below her, there was a sharp, steep sandbank where kingfishers had dug their burrows. Several of these raucous, blue-backed fishermen perched on the low, overhanging branches of the aspen poplars along the shore. As I watched from a small boat my thoughts drifted idly over the events of the previous two weeks.

We had arrived at Besnard Lake on May 14 just as the ice was breaking up. Most breeding eagles were already in the late stages of incubation. Huge ice pans were all around us as we crossed the two miles of open water to our cabin. Fortunately, a narrow channel remained open. For two days we watched as the ice was blown toward the southern end of the lake, piled up on shore, and started to melt. It was May 18 before we began our annual survey of the nests—all 250 miles of the shoreline had to be searched. Each day as we set out from our cabin to do the survey we had found Cindy perched on Kingfisher Island.

Cindy carried a yellow vinyl marker on her right wing and a green one on her left. They told us she had hatched four years earlier. We had put them on her when she was about ten weeks old and nearly ready to leave the nest. Several times since then, during the summers when she was two and three years old, we had seen her again as she wandered around Besnard and Nemeiben lakes. This year she appeared to be staking out a territory on a small group of islands close to our cabin.

For three hours I sat in my boat and fished, wondering at Cindy's patience. She perched quietly with her head angled down toward the water as she scanned for fish. From time to time she would look up and around, perhaps watching for other birds. Her sleek white head caught the sun. Like a beacon it stood out beside the dark spruce below and the blue sky above. Near four o'clock, ripples stirred on the lake to the southwest. A breeze was coming. At our cabin island behind me the aspen leaves began to quiver. Then the higher branches started swaying. Beneath me the swell of the waves gently rocked the boat. I zippered my jacket and settled back in my seat to enjoy the movement. Just as I did that, the breeze reached Cindy. She bent her head down and forward, lifted her wings, and was off.

With the breeze from behind, her big wings took her swiftly east over the southwest point of Robertson Island. I started the motor and followed. Cindy flapped on steadily, straight for a long stretch of southwest-facing shoreline. Here, the thick spruce forest rose like a wall or a cliff above the shore and diverted the wind to produce an updraft. Just above the trees, Cindy spread her sails and soared "wing-wide upon the air." I cut the motor and watched.

Soon Cindy was joined by another mature eagle. Together the two white-headed birds swept back and forth along the mile of windblown shoreline. The other bird, a male, probably from nest D a mile to the north, was a little smaller and tended to soar higher. Without flapping a wing, they coursed effortlessly back and forth. At intervals, gusts swept across the water and furrowed the waves with thousands of ripples. As each gust reached the shore the trees shook and the eagles were swept momentarily higher.

I was so caught up in the elegance of the birds that I did not notice how quickly my boat was being carried toward the rocky shore. Surprised at finding myself almost underneath the eagles, and not wanting to disturb them, I started the motor and boated farther away. Settling down with my binoculars, I soon lost myself once more in wonder at the ways of eagles in the air. At times Cindy hung almost motionless above a spot where the waves beat against the shore. At times she swept momentarily inland before returning to cruise once more above the trees along the water's edge. The tips of her primaries constantly felt the wind and made small adjustments; they spread out, closed in, then swept up a little to keep her positioned where she wanted to be. Her tail too was always changing, angling one way, then another, in small fluid movements. All too soon, I was again swept in toward the shore by the wind and the waves.

Once more I started the motor and backed out into the bay. I wanted to be as close to Cindy as possible, but I knew that learning Cindy's ways in the wild depended on my staying far enough away so that I didn't disturb her. When I was forced to start the motor yet a third time

I shook my head at the elements—why could they not leave me in peace? But of course that was impossible. To go where the eagles go, I would have to follow the wind and the waves. And then it struck me. The same waves that kept carrying me in to shore would also carry dead or dying fish, food that Cindy was looking for. I had been concentrating on the effect of the wind on Cindy but forgot for the moment that part of the reason she was there was to look for fish.

Several years earlier, my brother Peter had regularly put fish out on the north shore of our cabin island. One day he had come to me puzzled; the eagles found the fish only when the wind was from the north. Now, at last, I knew why.

*T*he wings of an eagle, six to seven feet from tip to tip fully stretched, are its ticket to the skies. Let us take a close look at the wide, long wings of a Bald Eagle and how they are designed to suit its needs. The humerus, the long bone in an eagle's wing that extends out from the shoulder, is hollow to conserve weight. Indeed, so carefully is the skeleton constructed that the bones of an eagle weigh, in total, less than half the weight of its feathers.[1] Extending beyond the humerus is the ulna and its associated bone, the radius. This middle part of the wing, where the sixteen secondary feathers are attached, provides a considerable proportion of the wing's total surface area. Beyond the ulna and the radius are the metacarpal

An adult Bald Eagle about to land (John E. Swedberg).

Facing into the wind, a young Bald Eagle holds out its long, wide wings to catch the breeze (Gary R. Bortolotti).

bones. Here on the outer third of the wing the ten primary feathers are attached; to some extent they can be adjusted individually, somewhat like fingers, so that as the bird flies it can manipulate them to reduce turbulence and increase its maneuverability or stability. The alulae, the small "bastard wings" at the wrist on the leading edge of the wings, are like the slots on the wings of an airplane and help to give control at low speeds.

A long, wide wing, the quintessential attribute of a large soaring raptor, is a hallmark of the Bald Eagle. It is this broad extended wing that allows it to soar well. Time and again I have marveled at the ability of a Bald Eagle to climb higher and higher, circling into the crisp blue of a spring or summer sky. The size of an eagle's wings relative to its weight allows an eagle to glide forward losing altitude only slowly. When an eagle finds a column of rising air (a thermal) or a region where air is forced upward against a cliff, then it can gain altitude because the surrounding air is rising faster than the eagle is losing height. If the air is absolutely still, with no movement up or down, eagles cannot soar. Indeed, this is why, on calm days, eagles usually perch.

The large wings, though advantageous for soaring, may make landing and takeoff difficult. The fluid grace of an adult coming into a perch is deceptive, for it comes only after long practice. Watching the awkwardness of a newly fledged eagle in the air gives a real appreciation of what a master adult flier has achieved. Arriving at or departing from an exposed perch in anything but a light breeze, an adult eagle, like an airplane, will almost invariably face into the wind.

An eaglet on its first few flights will sometimes try to land with the wind behind it. Several times I have seen a fledgling grasp momentarily at its intended perch before realizing that its forward momentum was too great to land successfully. Sometimes the bird manages to continue on its broken flight. Sometimes it is slowed enough by grasping at the branch that it falls; tumbling, the young eagle may manage to grasp a lower limb and hold on to perch there. Often the bird ends up on the ground. A particularly tenacious bird may succeed in grasping the branch, but unable to stop, it swings forward and comes to rest hanging upside down, wings extended and drooping, looking like a disjointed, overgrown bat.

At first, we presume that an eagle is an eagle is an eagle. That could not be farther from the truth, particularly when we are interested in how eagles fly. Remarkable changes, described in the last chapter, occur in both size and shape during at least the first four or five years of a Bald Eagle's life. Feathers of the wing and tail of fledgling eagles are longer than those of adults. As an eagle matures, its wings become shorter and narrower and its tail shorter with each successive molt.[2] The larger wing of an immature or subadult eagle gives it a lower wing loading (the ratio of the weight of the bird to the area of the wing). In other words, the more wing area the weight is distributed over, the easier it will be for the bird to be carried aloft in a thermal or updraft. There is also evidence to suggest that immatures weigh slightly less than adults (probably less muscle mass and fat), further aiding the young birds in obtaining lift.[3] The greater wing area of the immatures also permits them to fly slower and perhaps to soar in tighter circles (as needed in smaller thermals) than adults. The longer primaries of the immatures may also facilitate slower flight by allowing larger notches between the feathers at the tips of the wings, a feature that maintains lift while reducing drag.[4]

Rising air currents usually start when the sun heats the earth's surface in a differential fashion; for example, a large, flat, rocky surface may heat faster than the nearby forest, so the air above the rock is warmer relative to the air around it. The air above the rock then rises, and if the area is large enough, a thermal develops. Given its wing loading, an immature can rise up in smaller and weaker thermals than an adult. The adult eagle, however, is designed to glide farther and faster starting from the same height as an immature. This is a real advantage in cross-country flights, particularly with flights into the wind. Thus an adult that has a territory in summer and must go out from its nest and return (basically spending as much time going downwind as it does going upwind) is better equipped for the upwind leg of the trip than an immature would be. Unlike an adult, an immature does not need to feed young and is not tied to any specific area.

When soaring (below) an eagle's wings are stretched out to their fullest. When gliding (top) the wings are pulled inward somewhat and the tips are pointed backward (Naomi Gerrard).

Immatures are thus free to drift downwind and to wander across the countryside; in fact, that's what they do.

The differences in the size and shape of the wing between adult and immature Bald Eagles also give rise to other differences in their flight. For any bird there is a speed that is most efficient for steady flapping flight. This optimum speed depends on the size and shape of the wing and the weight of the bird. Observations on Besnard Lake suggest that adult eagles flapping steadily in calm air fly at 28 to 32 miles per hour. Immatures, with their relatively larger wings, flap slower than adults. That changes progressively as an eagle ages. At Besnard Lake, flapping rates of 13 immatures averaged 167 flaps per minute; of 2 near-adults, 177 flaps per minute; and of 28 adults, 188 flaps per minute.

Changes in wing shape with age are not the only important variation among eagles that affects flight. There are also sex differences, as female Bald Eagles are heavier than males. The wing length of females is also greater, but proportionately less so than weight. For example, the male of a mated pair at Besnard Lake weighed 3.9 kilograms (8.6 pounds), and his mate weighed 4.5 kilograms (10 pounds).[5] The wing area of the male was 5,600 square centimeters (870 square inches), that of the female 6,000 square centimeters (932 square inches). The wing loading (weight/wing area) of the male was therefore about 7 percent less than that of the female, and his weight was 15 percent less. The difference means that males can rise more quickly in thermals and soar or glide in the typically weak updrafts of the early and late portions of the day, when females are less suited to flight.

We have followed several individual eagles to try to understand the relationship between wind, time of day, and flight patterns. Our findings with one of these birds — Cindy of the introduction — are illustrative. When the wind was light, Cindy spent almost all her time perched. However, when the wind speed increased, as shown in the story at the beginning of the chapter, and gave rise to turbulence and updrafts, Cindy soared and glided proportionately more. That was particularly true during the middle of the day, when the effect of the sun was working as well to create the best thermals. For example, when the wind speed was high, more than 25 kilometers per hour (15.6 miles per hour), Cindy spent half her time soaring and gliding between 1:00 and 4:00 P.M. The proportion of time soaring and gliding decreased to 25 percent during the morning and later afternoon in the presence of similar winds. By comparison, when the wind speed was less than 15 kilometers per hour (9.4 miles per hour), then less than 2 percent of Cindy's time was spent soaring or gliding at any time of day. We were surprised at how dependent her soaring was on the availability of suitable winds. Smaller and lighter male eagles observed at the same time and later were less dependent on strong winds, soared more in light winds, and were more active in the early and late parts of the day.

Eagles are capable of sustained flapping flight, but they usually spend little time doing it. During the month when Cindy was observed intensively, she averaged less than two minutes per hour in flapping flight.[6] That is not surprising when one considers the large expenditure of energy required by the pectoral and supracoracoid muscles to power the huge wings. The energy needed to maintain a bird in flat soaring or gliding flight is much less, perhaps a twentieth or less the power needed for flapping.[7] Therefore, eagles will always choose to soar or glide when possible.

Efficient flight, always important to an eagle, is critical during migration. At no other time do eagles spend so long on the wing. It is perhaps here that we can best see how terrain and weather conditions influence flight. Three flight patterns occur most often (see Chapter 10).[8] Eagles may ascend in a thermal, then glide down, traversing the countryside as they do so. Such a glide may take an eagle several miles before it needs to find another thermal. The distance traveled depends on the altitude of the eagle, its wing loading, and the wind direction. Alternatively, eagles may circle steadily downwind using what is called a street of thermals, a straight series of rising air masses often resulting from a single powerful thermal source. And third, eagles may use the rising air generated when wind sweeps against a cliff or other raised feature of the terrain. Although they are generally reluctant to flap, eagles can occasionally be seen pumping their wings hard for considerable distances (for example, in bad fall weather when lakes freeze over and eagles must move south to find prey). Often an eagle flaps only at the beginning of a day's migration, to take it into its first thermal, and at the end of the day to get to a good roosting place.

Though we understand eagle flight better now than before, it is, after all, not the statistics that provide the allure of these birds. Bald Eagles may appear ponderous and labored when they are flapping, but it is their incredible ability to soar high and far that draws and holds our attention more than any other aspect of their flight. That ability is absolutely dependent on the weather, particularly on the wind and the sun. It is only when an eagle has learned how to use the everchanging pathways in the air that it can truly become a master flier.

4. Talons Awaiting

When he glares on his prey below.
Charles West Thompson

Jon —

It is five o'clock on a cool, clear, early morning in June when I step out of the cabin leaving my wife, Naomi, and our two children, Pauline and Charles, asleep. On my way I check the wind — south-southeast at eight to ten miles per hour — push a few pieces of bannock into a bag, load the boat, and set off.

It is 5:25 A.M. when I find Cindy, our four-year-old color-marked Bald Eagle, perched atop a spruce on an island southeast of her usual hunting spot. This morning she is fishing a channel between two islands. Waves cross the wide open lake to the south and sweep past below her. The sun is bright behind her as she gazes intently down and toward the south and west from her perch on the east side of the channel. For two hours she watches for fish, but finding nothing she returns to Kingfisher Island.

Her white head sparkles as the sun streams down on her; she preens for several minutes, arranging and oiling her feathers. Then, turning her attention once again to her search for fish, she visits several perches on Kingfisher Island and stays for a few minutes to half an hour on each. A little after ten o'clock an adult enters the eastern part of her territory and lands on a small island. Cindy, now on her crow's-nest perch, calls out — a high-pitched *whee-he-he-he* — and flies toward him. She lands on a high spruce not far away, leans over and down toward him, and calls again. The intruding adult, Yarak, a male from nest T (about four miles southeast) stares at Cindy; a few minutes later he flies off. East of the island Yarak catches a thermal, rises quickly up to two hundred feet, and circles. I watch. He suddenly plummets down. Yesterday, he caught a fish in front of Cindy and she chased him vigorously. Today, he descends behind an island, hidden from me, in his own territory. I presume he catches a fish and returns low to his nest, for I do not see him rise again.

Cindy continues her vigil for another ten minutes. Then she flies east to perch at the south end of the next island. Once again she has a good view of the oncoming waves. She scans them carefully, watching and waiting for fish. For fifty minutes she waits. Then she moves east to the next point of land. Again waves sweep in to Cindy across a large expanse of open water, but again they bring no fish.

The sky gradually fills with cumulus clouds, each one marking the top of a thermal. Shortly after one o'clock, a mottled brown immature eagle flies low between two islands and enters Cindy's territory. Cindy launches from her perch, flies rapidly after it, pumping her wings hard. She chases it off, then chooses a thermal and rises, circling easily and gracefully. Her wings are flat across, but her whole body is tilted slightly toward the center of the thermal as she sweeps around. At the tips of her wings her primaries curve slightly upward. With the southeast wind the thermal is not perfectly vertical but rather forms a cone angling northwest. Cindy drifts with the thermal, making a large sweep during the downwind (northwest) half of each circle and a foreshortened curve on the upwind (southeast) half. Across a large bay she goes, gaining altitude — 200, 300, 400, then 500 feet up — and moving farther away — now a mile from where she entered the thermal — until I can barely see her.

Just as it seems as if she might go on and up and away forever, she sets her wingtips slightly back and glides. It is a long gradual glide, first east, then south. She loses altitude slowly, passing high over her last perch as she makes her way along the shore, then out over several islands. For a mile and a half she cruises south, then, still 250 feet in the air, she cranes her neck to the west. Perhaps she has seen something on the water. She changes direction. In a slow arcing descent she covers another mile before reaching water level. Moments before she meets the surface, she swings her legs forward, brakes back slightly with her wings, and grasps a fish lengthwise. Her back talon hooks into the side of the fish while her others encircle it. With a strong flap, up and forward, she pulls it smoothly from the water.

Cindy, her feet barely wet, now flaps hard, rises gradually, and lands on a huge house-size boulder on the edge of the nearest island. She looks around. Then she digs into her prey, pulling off and devouring huge chunks of flesh with her big yellow bill. It is a reasonable-sized fish for her, about a pound. She works quickly. In four minutes she has entirely consumed it.

She pauses, content, wipes her bill on the rock, hesitates for a moment, looks around, and then glides off the rock using the shoreline updrafts. Soon she lands, gracefully, on a spruce top at the far corner of the island. Her day's search for food has taken almost eight hours, most of it as a patient, perching fisherman, but with ten minutes of high soaring flight and a three-minute downward glide.

A Bald Eagle catches a fish (John E. Swedberg).

The diet of the Bald Eagle can be extremely variable. In much of its range and during much of the year, fish is an important and often the major component of an eagle's diet. The species of fish taken vary considerably from one locale to another. In north central Florida, three quarters of the prey items found in or near eagle nests were fish, and more than half of the fish were brown bullheads.[1] The size of the bullheads taken ranged from half a pound to one and a half pounds. Blue and white catfish and lake chubsucker were also frequent prey. Of the nonfish food items, American Coots were common (11 percent of all items). Mammals were uncommon prey, with only a few water rat and rabbit remains found. Similarly, three quarters of food items found in interior Maine and two thirds of those on Cape Breton Island, Nova Scotia, were fish.[2] Almost three quarters of the fish in Cape Breton were cod; in Maine brown bullheads, white sucker, and pickerel were all common. Curiously, Great Blue Herons constituted 9 percent and snowshoe hares 5 percent of prey items in Cape Breton.

In most parts of northern Saskatchewan eagles feed almost exclusively on fish during the breeding season. Cisco (also called tulibee or lake herring, a relative of the whitefish), suckers, burbot (or ling), northern pike, and walleye are the principal species caught. Many of the fish may be picked up dead; however, the large preponderance of cisco, a surface-feeding fish, over whitefish, a more bottom-feeding fish, suggests that surface feeders are particularly susceptible to predation by eagles (see Appendix 3).[3] Species likely to be found in shallow water, such as sucker and pike, are also common prey. It is interesting that the distribution of prey species varies from nest to nest. Some eagles may specialize in one type of fish (e.g., burbot) either because it is more available in that part of the lake or because they have become particularly adept at catching that species.

Bald Eagles use a number of techniques to search for and catch prey. In northern Saskatchewan, as the account at the beginning of this chapter suggests, eagles spend a great deal of time perching. When a fish is sighted, the eagle will usually descend quickly and smoothly from its perch. At the water's surface, the eagle will reach a foot (or both feet, for a big fish) down to scoop up the fish and carry it off. The location of fishing perches is chosen to take advantage of the incoming waves, which might bring dead or dying fish. Gliding along shorelines where the wind creates updrafts gives the bird a similar advantage in finding and catching dead, sick, or injured fish, as shown in the description at the beginning of Chapter 3.[4] When fishing on the wing, eagles may course low over the water or soar high. To descend quickly from a height to catch a fish, eagles sometimes pull in their wings and stoop down, dropping like a stone, but more commonly they will descend in tight circles or tilt their wings vertically, sideslipping to lose altitude.

Immatures and adults tend to have different foraging styles. Immatures, as a result of their larger wings and tail, are searchers, or wanderers, looking for carrion; adults are attackers, more often hunting live prey in a localized breeding or wintering territory. Such was the case for the eagles of Besnard Lake. Observations of live (still wriggling) fish being brought to nests suggest that territorial adults catch many live fish; immatures, in contrast, appear to feed primarily on carrion. The importance of floating, dead, or injured fish to immatures is suggested by the finding that the number of such eagles on Besnard Lake correlates with the availability of dead floating fish. Another example is from Amchitka Island, Alaska, where the immatures in their first three years feed heavily on carrion—harbor seals, sea lions, sea otters, fish, and even whales. Steve Sherrod and his colleagues made observations suggesting that the adults on Amchitka were better than the immatures at catching prey, including live sea otter pups, and relied less on carrion.[5] For example, during their study two beaked whales washed up along the coast—the first attracted 55 eagles and the second 65. The wandering immatures were more likely to take advantage of this food; almost three quarters of the eagles at the two carcasses were immature even though immatures made up only about a third of the population. Al Harmata in the San Luis Valley of Colorado made further observations that suggested that immature searchers hunt more on the wing (soaring) and that adult attackers, which are less efficient at soaring in weak thermals, hunt more from perches.[6]

We were curious whether eagles more easily recognize and catch a fish that is right side up (dark side up) or upside down (white side up). We found that white side up fish were much more quickly spotted and caught. To our eyes, and apparently also to the eagles', the white belly was more visible than the dark back. Territorial adults were the subjects of our experiments, showing that even breeding adults quickly recognize and utilize dead fish when they are available.

When actively looking for dead or injured fish, eagles watch for clues to where there may be fish as much as for the fish themselves. A gull at the edge of the lake or on the open water will be watched closely by any nearby eagles. If the eagle suspects that the gull has found a fish, it will usually fly over to check things out and often descend to take the fish for itself. On the wintering grounds in Colorado and elsewhere, ravens, crows, magpies, and other birds take the place of gulls. Indeed, Harmata found that animal carcasses not visited by those corvids were rarely visited by eagles, but that eagles often arrived shortly after a crow or a magpie.[6] Some Bald Eagles also appeared to follow Golden Eagles (which frequently killed jackrabbits and ducks in the San Luis Valley). At a carcass, the Bald Eagles never attempted to displace the Golden Eagle but fought among themselves after the Golden Eagle had left.

In a similar vein Susan and Richard Knight, in studies along the Nooksack River in Washington, found that Bald Eagles were more likely to go to feed at a fish if another Bald Eagle was already there.[7] A flying or perched eagle dropping suddenly to a carcass may serve as a tip-off; in Colorado,

An immature eagle beside a deer carcass in winter (John E. Swedberg).

eagles that descended rapidly to a carcass were often joined by up to twelve others within five minutes.[6] Farther from the food source, birds may use additional clues. Spiraling columns of soaring eagles will gather above the site. Riley McClelland calculated that such a group of spiraling eagles might be visible, and could serve as a signal, to other birds as far away as fourteen miles or more.[8]

Fish spawning in large numbers in shallow water are vulnerable to predation by eagles. In northern Saskatchewan from mid-April to mid-June eagles gather to feed at spawning streams that are full of suckers, pike, and walleye. A good description of eagles feeding on mullet in November 1936, when the fish were schooling in great masses to spawn, was contributed by Edward Reimann.

> I happened upon one of these schools and saw six Bald Eagles [four adults and two immatures] . . . fishing for themselves. . . . The eagles circled directly over the school, about fifty feet above the surface of the water. One would break away, go about one hundred yards off, and then start out at full speed toward the school, gaining momentum and setting its wings in a long diagonal glide down to the surface. It would then reach down into the water, immerse one leg and scoop out a fish, never stopping for a second. . . . Only one bird was seen doing this at one time, while the others would wait until one made a catch and flew away; then another would go through the same procedure.[9]

Some breeding Bald Eagles live primarily on food other than fish. Eagles nesting adjacent to the Blackwater Marshes of Maryland feed primarily

The female Bald Eagle (on the nest) has just delivered a fish, and her two chicks huddle over and inspect it (Gary R. Bortolotti).

on muskrats and waterfowl.[10] In the extensive marshes just south of Cumberland Lake in northern Saskatchewan, there are vast numbers of breeding waterfowl; here Bald Eagles eat almost exclusively ducks and other water birds. Where eagles nest adjacent to colonies of seabirds, eagles will subsist predominantly on them. For example, food studies in coastal Maine showed that only 17 percent of prey items were fish, but 76 percent were birds, particularly Black Ducks and colonial nesting species like gulls and Double-crested Cormorants.[2] However, fish were likely under-represented in the sample; it was based on prey remains, and frequently little is left in the way of evidence from a fish meal. Fine-boned fish like the American eel and the tomcod were commonly observed to be caught but were nearly absent in the food remains found in or below the nests. Similar results were found on Amchitka Island, where 61 percent of prey items were birds, but as in coastal Maine, fish remains were believed to be underrepresented.

Sometimes eagles use unusual methods to obtain a meal. On Petrel Island in Alaska, eagles regularly excavate storm petrels, murres, auklets, and puffins from their burrows. However, it appears more usual for eagles to catch injured or sick seabirds.[11] Sometimes an eagle that catches a sizable seabird, such as a cormorant, will have unexpected difficulties, as illustrated by the following account written by a Mr. Worthington in Darien, Georgia.

> The other day I noticed a Bald Eagle hovering over the sound, much the same as a fish hawk does when about to strike a fish. Suddenly he plunged down and grappled with what I supposed to be a large fish but was unable to raise it from the water and after struggling awhile he lay with wings extended and apparently exhausted. After resting a minute or two he again raised himself out of the water and I saw he had some large black object in the grasp of one of his talons, which he succeeded in towing along the top of the water toward the shore a short distance, and then letting go his hold. He was then joined by two other Eagles, and by taking turns they soon succeeded in getting it to the shore. Investigation proved it to be a large Florida cormorant.[12]

The story also illustrates the upper limit of what the smaller Bald Eagles of Florida can carry, probably about four pounds, the weight of a large Double-crested Cormorant. In British Columbia, Wayne Campbell observed an eagle that, similar to the bird in Florida, had difficulty lifting a cormorant from the water; however, after swimming with it to a small rock islet, the eagle flew "laboriously" to another island three miles away.[13]

On the San Juan Islands of Washington State, the most common food of nesting eagles is the European hare, introduced for the commercial market and then released in numbers when the market declined.[14] The hares are abundant, and road-killed or hunter-injured animals are readily available.

In fall and winter the diet of Bald Eagles tends to be more varied than in summer, although fish are still important. During the migration south, ducks and geese crippled during the hunting season are an important food for interior eagles from Ontario, Manitoba, and Saskatchewan. As the lakes freeze on the Canadian prairies, hunter-crippled waterfowl become trapped in small openings in the ice and fall easy prey to eagles. The peak of eagle migration occurs at the time of freeze-up, perhaps to take advantage of this food source.[15] In many parts of the American west eagles will congregate at refuges where geese are wintering and feed on them as well as on local fish. On occasion, Bald Eagles will prey on swans, as noted by Larry Dalton at Desert Lake in Utah.[16] A small group of thirteen Whistling Swans were killed at the rate of almost one a day in late December and early January. In the one kill observed, the eagle flew low over the ground, then dropped behind a dike where two swans were resting. One flew away, the other was caught by the eagle after flying about eight feet.

In parts of Utah, Colorado, and Oklahoma, Bald Eagles prey on the abundant jackrabbits. A notable location is near Provo, Utah, where Clyde Edwards described communal roosts of Bald Eagles in the mountain canyons. During the day these eagles descend to the valleys to hunt, often catching hunter-injured animals.[17]

Eagles in many areas have been reported feeding on the carcasses of

A 2½-year-old Bald Eagle (known age) feeding on carrion in winter in Maine (Mark A. McCollough).

dead sheep or other large mammals, usually animals that have died natural deaths. During the hunting season, hunters frequently eviscerate deer in the field. Eagles can be found feasting on such gut piles. To what extent Bald Eagles kill healthy piglets, lambs, and kid goats has been widely debated. As early as 1709, John Lawson reported that an eagle had carried off young pigs and noted that one poor pig made "such a noise overhead that strangers have thought there were flying sows and pigs."[18] However, recent studies indicate such predation is very rare, though it may happen when a young animal is not closely attended by its parent.

The techniques used by eagles (and best developed in the "attacking" adults) for catching live prey other than fish depend on the situation. Weakened, injured, or just slow animals tend to be captured most often. The following are three observations made by Dick Dekker in Alberta.

> 6 November 1976, 1100 h, Hastings Lake. Flying at about 10 m [meters] against a strong wind over the frozen lake, an adult Bald Eagle approached a water hole in which about 100 ducks, mostly Lesser Scaup were massed together and splashing about. A lone female Mallard flushed well ahead of the eagle and dropped back into the water just after the eagle had passed. In a very swift plunge the eagle doubled back, seized the Mallard and carried it to the ice.
>
> 6 November 1976, 1300 h, Beaverhill Lake. An adult Bald Eagle flew up from the ice and approached a flock of about 60 Mallards swimming in a

patch of open water. All except one of the ducks flushed and left. The eagle turned to the lone Mallard, seized it and carried it to the ice.

14 November 1976, 1230 h, Cooking Lake. An adult Bald Eagle that had been standing on the ice for some time flew at 2-3 m to a water hole. About 80 ducks, mostly Lesser Scaup, hurriedly left the edge of the ice, on which they had been resting, and entered the water. The last bird, a drake scaup, was seized by the eagle in one foot and carried away.[19]

The following observation illustrates the technique of an adult Bald Eagle that regularly preyed on water birds, in this case coots. The bird came daily and perched on the top of a tree near a river where large numbers of coots were feeding.

At the first sight of the Eagle the Coots all huddled together, remaining so during his rest, swimming about aimlessly and casting uneasy glances up in the direction of their enemy. The moment the Eagle lifted himself from his perch, the Coots seemed to press towards a common centre until they were packed so closely together that they had the appearance of a large black mantle upon the water; they remained in this position until the Eagle made his first swoop, when they arose as one bird, making a great noise with their wings, and disturbance with their feet which continued to touch the water for the first fifty or one hundred feet of their flight. This seemed to disconcert the Eagle who would rise in the air only to renew his attack with great vigor.

These maneuvers were kept up, the Eagle repeating his attack with marvelous rapidity, until, in the excitement and hurry of flight, three or four Coots got separated from the main body; this circumstance the Eagle was quick to discover and take advantage of; it was now easy work to single out his victim, but usually long and hard to finally secure it. I have never seen him leave the field of battle, however, without a trophy of his prowess, though I have seen him so baffled in his first attempt to separate the birds, that he was compelled to seek his tree again to rest.

On one occasion, after separating his bird from the flock, he spent some minutes in its capture—the Coot eluding him by diving; this frequent rebuff seemed to provoke the Eagle to such an extent that he finally followed it under the water—remaining some seconds—so long, indeed, that I thought him drowned; he finally appeared, however, with the bird in his talons, but so weak and exhausted that he could scarcely raise himself above the water, and for the first thirty or forty yards of his flight his wings broke the surface of the water; very slowly he made his way to the nearest tree, where he alighted, on the lowest limb, to recover his spent strength.[20]

On occasion, eagles will team up to catch healthy prey. The behavior of a pair of eagles that nested in a large live oak tree a mile and a half back from the beach along the Gulf of Mexico in Louisiana in the 1890s was described by E. McIlhenny.[21] At that time three quarters of the hundreds of thousands of Snow Geese in North America wintered within seventy miles of the site.

A very large flock of geese was feeding on the marsh on both sides of the bayou, only a short distance from camp. I was sitting outside the camp at about eleven o'clock, when my attention was attracted to two Bald Eagles coming from the direction of their nest which could be plainly seen from where I sat. As they neared the flock of geese which were sitting quietly on the marsh and beach, one of the eagles, the larger one, which I learned afterwards was the female, started upward circling. The other eagle flying slowly came to the flock of geese and swooped down as if to make a strike. A lot of the geese toward which he was darting rose and changed their position flying to one side or the other. He flew among them, and I expected to see him make a kill. Instead, he turned and came back swooping at the main flock sitting at the ground. All this time his mate was slowly circling on the sea side of the flock. [As we shall see, this was probably important so that the geese that flew up went inland.] After the smaller eagle had made five or six dives at the flock on the ground, he got under the geese that had become frightened and began to climb. Instantly the larger eagle swooped with a rush, joined her mate and both started circling under the rising geese who were going up and inland as fast as they could. The smaller eagle circled slowly under the geese, the large one very rapidly around them gaining height fast in large circles. When the geese were about five hundred feet in the air, one of them left the other two. Then the smaller eagle narrowed its circles under the goose driving it rapidly higher. Suddenly the larger eagle which had gotten considerably higher than the goose, shut its wings and shot with great speed down at it, struck the goose fairly and with bowed wings flew to its nest, its mate following. The eagle carrying the goose did not flap its wings after the strike until obliged to check itself to alight on its nest. The smaller eagle did not attempt to catch a goose, then or at any other time, and I saw the same performance every day while I was in camp and always about the same time each day. I visited this eagle nest while I was in camp and counted the heads of thirty-one Blue Geese, fourteen Mallard ducks and seven Pintail ducks. The young eagles were then about three weeks old.

Another example of cooperative hunting was given by Clyde Edwards, observing in winter near Provo, Utah, where Bald Eagles were hunting rabbits in an arid intermountain valley.

Hunting techniques consist of short coursing flights [at 3 to 10 feet above ground] back and forth over vegetative cover concealing prey. There is much intermittent perching on any available structure, and since several birds are in the vicinity, some will be sitting and others flying. It appears that the perching behavior is as much a flushing technique for concealed prey as is the low coursing flight. A rabbit will bolt from cover as frequently from a bird landing near it as from one flying low overhead.

I have observed eagles to land and then walk along on the ground through low brush in what appears to be deliberate attempts to flush prey. This method is apparently quite successful because airborne birds indicate by their activities that they see prey and kills [are] subsequently made.[17]

Many observers have noted that eagles rob Ospreys. The Osprey catches fish in a very different manner than the Bald Eagle, spending much of its time in flapping flight and frequently hovering (which eagles almost never do). When an Osprey dives, it usually plunges into the water, spray erupting around it, and often submerges in its attempt to catch the fish. Eagles usually descend more gradually and pluck a fish from the surface, getting no more than their feet wet. Ospreys tend to fish along the calm side of the lake or over open water, the portions of the lake less used by eagles. To see an eagle take a fish from an Osprey is a real thrill. G.B. describes one encounter at Besnard Lake: "As we boated across the lake, an Osprey flew low directly over our heads carrying a fish ten to twelve inches long. This was an uncommon sighting; no 'fish hawks' breed on the lake. As we started to record the time and location of the observation in our notebook, a shadow passed suddenly over us. It was an adult Bald Eagle flapping so hard that we could hear its wingbeats. In a matter of seconds, the pursuer was upon his target. The size difference between the two birds was astonishing. The eagle attacked from a few feet above and to one side of the Osprey — once, twice, and then a third time. On each pass the Osprey visibly flinched, for the eagle's talons barely cleared its back. We anxiously awaited the outcome. Many observers had described how typically the Osprey drops its fish, and with surprising agility the Bald Eagle snatches the fallen prey in midair. We were not prepared for what happened. After three unsuccessful attacks, the eagle turned to brute force. This time coming up fast from behind and below, the eagle flipped onto its back, thrust its talons upward, and ripped the fish right out of the Osprey's grasp. What a sight! After quickly righting itself, the eagle turned and flapped leisurely to deposit the booty on its nest. Where the Osprey went we do not know, but it was the last of the species to be seen that summer on Besnard."

On another memorable occasion, this time in Florida, G.B. watched an immature eagle chase an Osprey. The aerial acrobatics were impressive. Wings beat furiously and bodies twisted and turned as the two birds zigzagged across the countryside. A second immature eagle soon joined in. However, at some point in the exhausting chase the fish was dropped. As the chase was over land, the prey was lost in the vegetation and never recovered. One cannot help but wonder whether the stolen food could possibly replenish the energy expended in all that flapping. Perhaps eagles sometimes don't know when to quit.

5. To Find an Eagle and Its Nest

Bird of the broad and sweeping wing!
.   .   .   .

The skies, thy dwellings are.
James G. Percival

Jon—

Chesapeake Bay. It is 6:30 A.M. on a bright mid-May morning as I step out of my hotel in the middle of Washington, D.C. Hard as it is to believe, I am shortly to be taken to an active Bald Eagle nest less than ten miles outside the city. My host, Jackson Abbott, a seasoned eagle watcher, began to study the Chesapeake eagles in the fifties. In the early sixties he parlayed his connections with the military into the first aerial survey of the Chesapeake region. (The pilots from Fort Belvoir needed to accumulate flying hours, and looking for eagles and their nests could be done at the same time.) At Great Falls, along the Potomac River northwest of Washington, we leave the car and walk across a small canal to a viewing area. Below and to the left are rapids. Upstream is a long island. On the edge of the island, amid the lush green deciduous trees—a softer, gentler, mistier hue than the harsh dark green of the northern spruce forest—is an eagle's nest. We walk along the canal, past two families of Canada Geese, and then through the forest to the river's edge for a better look. The nest is high in a dead tree. One of the adults is perched on a limb above the nest. Between us and the nest the swift shallow waters run quickly over numerous small rocks; many smooth rounded tops show just out of the water. As we watch, a second adult arrives and flies down to land gracefully on one of the rocks. Not used to the scenery, I blink. It seems like a green fairyland with a magical river running by.

Northwestern Ontario. I sit on the riverbank waiting my turn, passing the time watching goldeneyes courting on a morning in late April. A male idles up to a female and stretches out its neck. Its black and white back and the iridescent green sheen on its head sparkle, beautiful in the sunlight against the backdrop of the deep blue water and the crisp white ice on the bank behind. Another male swims up; the first leaves the female and charges the competition, his bright orange legs paddling hard. The intruder retreats, tries again and again, but eventually leaves— sulking, it seems to me. I continue watching. Soon the Super Cub returns, lands smoothly on the water surface, and taxies to a stop beside me. After refueling, we are off into the clear blue. Below us stretches the snow-covered ice on Lac Seul. We cross to the south side of the lake, where a small square is marked on the map. Our task is to survey all the shoreline within this section. At the border of the square, I motion to the pilot to go lower until we are just above the tops of the aspen and pine. Slowly and carefully we follow along the shore; the treetops pass quickly just beside us. I crane my neck, alternately looking in front, to the side and behind. Ahead, atop a spruce, a white spot gleams. It is a mature Bald Eagle. We pass it without seeing a nest. I motion to the pilot to circle, and as we come around I see a light brown bundle of sticks on the side of a pine tree. An adult eagle lies low (incubating eggs) across the center. I mark it on the map. We continue.

Arizona. It is early December, and Teryl Grubb describes one of the nests he has studied, "On a little pinnacle that caps a rock protrusion jutting out into a hairpin bend of the Salt River is a nest site that has been in use since the thirties." We continue in a jeep to bump our way over rough, rocky roads through a cactus and mesquite forest. Earlier that day, we visited a nest on the side of a cliff along the channel of the Verde River a short distance downstream from the Bartlett Dam. Tall saguaro cacti and the occasional thin-leaved mesquite tree were present along the river and the surrounding hills, but they were hardly appropriate for an eagle's nest. In this dry region, suitable trees are rare; an old cottonwood was used in 1977, but grazing cattle effectively prevent any new cotton-woods from starting. In recent years the eagles had used the cliff nest.

As we come around the back side of Horseshoe Lake, another reservoir, I can at last see a few willows along the river and where the river enters the lake that might, in a pinch, serve for an eagle's nest. Even these are barely suitable. Teryl points out a metal tripod he erected nearby when an earlier nest in the best of these willows had fallen. To my left the sheer face of a flat-topped mountain rises about five hundred feet above the river and reservoir below. Teryl points up to it. "There is a small crevice about halfway up the cliff face that serves as an alternate nest site. It is only about eighteen inches across though, and really not quite big enough. That is why we put up the nest on the tripod." I gaze upward and scan the cliff with my binoculars. Two eagles are perched side by side just south of the nest.

The search for nests of the Bald Eagle involves visits to many of the wild and wonderful places in North America. Since the coming of

A nest in a poplar tree in northern Saskatchewan shows how close to the water's edge eagles may build (Jon M. Gerrard).

Europeans to this continent, the habitat used by these birds has shrunk until the majority of nesting pairs now live in wilderness areas. Even when eagles breed in relatively populated states, they tend to choose environs where there are the fewest people.[1] In the Chesapeake Bay region, for example, eagles are more likely to nest in habitat distant from human activity. In Manitoba, nesting eagles avoid sections of lakes and rivers where rows of cabins have been built, and so restrict their breeding to the more remote locations around lakes where there has been substantial development. A successful eagle's nest right beside a cabin in northwestern Ontario and another adjacent to a home in Florida provide notable exceptions to the general rule that eagles avoid locations with significant human activity. Nevertheless, at present, on a continentwide basis, rule number one for finding an eagle nest is to go to a wilderness, or at least to a relatively remote area.

Another important consideration in the search for breeding eagles is to look in the vicinity of a body of water. Bald Eagles, without exception, choose to nest within a reasonable distance of water. The need for both water habitat (for fish or other prey) and forest habitat (for nesting) has meant that eagles tend to concentrate along the shorelines of rivers and lakes. It is in fact typical of a species that uses two habitats that its population size is proportional to the length of the interface between the two—

the shoreline in the case of the Bald Eagle.[2] In most populations (for example, southeastern Alaska, Washington, and Besnard Lake, Saskatchewan) the average distance from a nest to water is less than a quarter mile. In some areas, like the Chippewa National Forest in Minnesota, there is a tendency for eagles to nest farther from water.[3] In north central Florida, nests even averaged more than half a mile from water.[3] In those areas greater distances may have resulted from increased human waterfront development. Nevertheless, even including those regions, nests farther than two miles from water almost never occur.

Around any lake or river, eagles are particular as to where they locate their nests. Local areas where food is more abundant or reliable are favored. Nests in northern Saskatchewan, for example, are more likely to be close to a good fishing area, particularly near a rapids where the water is free of ice early in the year. Similarly, in the Yellowstone region of the United States, eagles may pick areas near a part of a lake or river where the ice leaves early so that they can find food in the weeks between their arrival in early spring and the melting of the ice on the main part of the lake.[4] The farther north (or the higher the altitude, in the case of the Yellowstone region) the more critical is the availability of open water in spring. Large areas of the lakeshores in the northeastern corner of Saskatchewan are unused by breeding eagles because the ice thaws so late. In that region, only near rapids or fast-flowing streams can eagles be assured of open water in late April or early May, and only near such locations will nests be found.

The availability of suitable nest trees (or cliffs in some areas) also limits eagles' choices of breeding sites. Trees are important both for the nest itself and for fishing perches. For the most part eagles pick one of the taller stands of trees. Though they do not necessarily choose the tallest tree for their nest, it is usually one of the stoutest or sturdiest available.[5] Eagles tend to use trees with foliage that is not too dense so that there is plenty of room for flying to and from their nests. Perhaps for the same reason, eagles often choose a tree along the edge of a habitat — a shoreline or, as in the Chippewa National Forest, on the edge of a recently logged area.

In Arizona and elsewhere, in the absence of suitable trees, eagles will use cliffs, including a shelf on a sheer rock face more typical of the nest site of a Golden Eagle. Cliffs are also used where the trees are too small because of a subarctic climate or following a recent forest fire. In Saskatchewan, such nests tend to be on rounded, rocky hills near the water rather than on the steep rugged cliffs chosen by Bald Eagles in Arizona or used by Golden Eagles in the same part of Saskatchewan. Along the coast of treeless Amchitka Island in Alaska, eagles most frequently use sea stacks, cornices of rock that jut out into the ocean at high tide. Rarely, in regions where there are no appropriate trees or cliffs, eagles will nest on the ground. Once, in Michigan, a pair of eagles built a nest on the ground even when there were lots of usable trees nearby.[6]

A Bald Eagle's nest on a rock pinnacle, Salt River canyon, Arizona (Teryl G. Grubb).

To provide a look at the variety of habitats used by eagles, twelve different regions of North America are highlighted. For each, the known history of Bald Eagle use, the present distribution of eagles, and unique or characteristic aspects of their natural history are mentioned.

The Atlantic Coast

Florida

Over the years, the eagles of Florida have been studied by a number of gifted naturalists.[7] From 1935 to 1971 J. C. Howell kept track of eagles nesting in 24 territories on or near Merritt Island beside what is now the NASA base at Cape Canaveral; during that period the number of pairs of eagles occupying nests decreased from twenty to four. Charles Broley, as described in Chapter 1, banded eagles on the Gulf coast in the Tampa Bay region and south to Fort Myers from 1939 to 1959. We now know that DDT was a major factor in the decline of that population; however, tremendous changes also occurred in the coastal habitat, with thousands of acres cleared of the large pines (in which eagles prefer to nest) to make way for market gardening, housing, and other developments. In 1942 Broley knew of twelve active nests on Pine Island near Fort Myers "in a tract one and one-half by six miles. . . . In January, 1949, I was dismayed to find practically all the large timber had been cut and that only four nests were occupied! Many adult eagles could be seen, but they were just loafing around, loath to leave their old nesting territory, and yet unable to find a suitable nesting site."

In the years since 1972, when DDT use was curtailed, the population of eagles in Florida has increased, and by 1980 it had reached approximately 1,400 birds. There are still many eagles nesting in the Everglades and in north central Florida, though in other areas, such as the Tampa Bay and Sarasota regions, there are remarkably fewer than there once were; there simply is not enough suitable habitat left. Fortunately, some eagles have adapted to human development; since 1966, a pair of eagles has nested on the Kennedy Space Center a few yards from a heavily traveled highway and only a few miles from the launch pad for NASA's space shuttles.

Unusual observations in Florida by several people have contributed to eagle lore; these include records of an eagle that captured a tern carrying a fish, an incubating eagle killed by lightning, one eagle that incubated a Great Horned Owl's egg, and another that incubated a rubber ball. Charles Broley kept a record of the odd objects he found in Florida nests: a light bulb, a bleach bottle, a newspaper, shoes, a skirt, and even a pair of lace-trimmed, pink step-ins (panties). Certain of the objects, such as a bleach bottle floating in the water, might have been mistaken for prey (indeed, we have seen a plastic bottle pierced by eagle talons on Besnard Lake, probably because the white bottle looks from a distance like a dead fish floating belly up), but others may have been taken to the nest out of curiosity, for play, or for unknown reasons.

Chesapeake Bay
In the center of the east coast of the United States is a huge brackish sea spanning 3,200 square miles; its outlet to the Atlantic Ocean is but 12.5 miles across. Into this sea comes the outflow from 48 rivers and countless creeks. Along its 5,620 miles of shoreline, there was said to be, in 1890, one eagle nest for every mile.[8] The proportion of those nests occupied by breeding eagles is unknown. Even if it was only 20 percent, there would have been more than 1,000 pairs; if, as is more likely, 50 percent were occupied, there would have been more than 2,500 pairs of eagles. From current studies we know that when we include nonbreeding eagles, the total population is usually three to five eagles per breeding pair, or somewhere between 3,000 and 12,000 eagles in the Chesapeake in 1890. Forty-six years later, in 1936, from studies by Bryant Tyrell,[8] it was estimated that the region held but 600 to 800 nesting pairs. Both of these early estimates were crude, and

it is uncertain whether there was any real change over the period. What is clear, however, is that in the 26 years from 1936 to 1962, there was a precipitous decrease in eagle numbers. A census in 1962 of half the Chesapeake Bay found only 28 active nests. Five of the nests produced seven young. The rest failed. From that data, a conservative estimate would be that the total eagle population on the Chesapeake had decreased from about 3,000 eagles in 1936 to about 300 in 1962, with the number of young raised dropping off from 700 to 20. Gradually, since 1962, Bald Eagle reproduction and numbers have improved. By 1986 the number of active nests in the Chesapeake Bay had increased to 133, and 188 young were raised that year. Now eagles nest within a few miles of the nation's capital.

In the Chesapeake, eagles usually start building nests between November and January, carrying sticks and broom grass to a suitable crotch in a stately pine, oak, or tulip tree. In the good years, the thirties, eagles could harvest so much unpolluted food from the bay and the adjacent extensive freshwater marshes — fish, turtles, and even snakes — that 40 percent of nests had three eggs and 2 percent had four, making them perhaps the most productive eagles on the continent.

Newfoundland, Nova Scotia, New Brunswick, and Maine
Bald Eagles nest along the northeastern coasts of North America from the northern part of the island of Newfoundland to the southern part of Maine. The population of Bald Eagles in Newfoundland has been relatively little studied, but those in Nova Scotia, New Brunswick, and Maine have been thoroughly investigated, with 106, 12, and 64 breeding pairs respectively in the three regions in 1980 (1981 data used for Maine).[9] The Bras d'Or Lake region of Cape Breton Island in Nova Scotia has a fairly dense cluster of nesting birds. Many of them appear to stay in Nova Scotia year-round, though some, perhaps particularly the immatures, move south into Maine in winter. In Maine and adjacent New Brunswick, eagles are found both on the coast and along inland rivers. The birds of Maine were severely affected by DDT; only recently, with management programs including an eagle transplantation effort and a program to provide clean (i.e., free of pollutants) carrion during the winter (see Chapter 12), are the eagles doing well.

South Central United States

At one time Bald Eagles nested all along many of the rivers and lakes in the south central United States, including much of Texas, Oklahoma, Arkansas, Mississippi, Louisiana, Tennessee, and Kentucky. Eagles were still present and fairly widely distributed in the region up to at least the thirties. A. F. Gainer, writing in 1933, provided a short description of visits to a number of their nests — an active nest seen in 1899 in the topmost branches of a great cypress tree on Cypress Lake, just north of Vicksburg,

Mississippi; visits between 1919 and 1923 to three nests in cypress trees at Reelfoot Lake, Tennessee; and a visit in 1930 to four nests in giant white oak or elm trees in Horseshoe Lake, about twenty miles southwest of Memphis.[10] In the early thirties A. M. Bailey commented, "As hunters try to kill eagles at every opportunity, the birds are becoming rare."[11] By the early seventies Bald Eagles had been virtually eliminated as a breeding species from the entire area. The population has survived only in the dense cypress swamps of Louisiana just inland from the Gulf of Mexico (six young were raised at seven active nests in 1972) and in coastal Texas (four young raised at four nests in 1972).[12]

W. Dubuc in the seventies found that the usual diet of the southern Louisiana eagles was fish, especially carp and bowfin, with duck feathers found at one nest and a snake seen carried to another.[13] However, McIlhenny, who lived on Avery Island along the Louisiana Gulf coast, made many observations in the late 1880s and early 1900s suggesting that some pairs then specialized in waterfowl.[14]

Arizona

The present eagles may be the remnant of a population that was once more widely distributed, as shown by reports that Bald Eagles were formerly common residents in the White Mountains. Nesting along the Salt River in south central Arizona was documented as early as 1911.[15] Today, small numbers of Bald Eagles breed in trees or cliffs along the hot, dry valleys of the Salt and Verde rivers in central Arizona. From 1975 to 1979 six to ten breeding pairs constituted the entire population. By 1986 there were eighteen occupied nest sites, and seventeen young fledged from twelve of them.

Courtship begins as early as late September and continues until the eggs are laid between mid-January and early February. The young fledge between mid-May and mid-June. The latest young to fledge appear to be at the most risk of heat stress in the poorly sheltered nests in this land where the temperature may reach 110°F. The primary prey species are carp and channel catfish, with suckers, crappies, largemouth bass, waterfowl, reptiles, small mammals, and carrion also eaten to some extent.

Yellowstone National Park and Adjacent Area

On the high volcanic plateaus of Yellowstone National Park and the surrounding region (together called the Greater Yellowstone Ecosystem) we can see in microcosm the story of the Bald Eagle in the western states during the last century.[16] Outside the park, the number of Bald Eagles nesting along suitable rivers and lakes was, by 1960, probably far below that of the mid-nineteenth century. Fortunately, protection given within the park resulted in continued eagle nesting. Indeed, the number of nesting Bald Eagles in Yellowstone has changed little since 1872, when it was established as the first American national park. Even the locations of individual territories appear to have changed little. There are, though, a few telling exceptions. The originally barren Shoshone and Lewis lakes were stocked with fish, and nests along their shores today testify to the beneficial effect of human intrusion. On the other hand, a site at Grant Village was abandoned by eagles when construction began nearby in 1963, and a territory near Fishing Bridge (which is heavily used by people) has had little eagle activity since 1973. From the mid-sixties to the mid-seventies a low rate of reproduction was probably due to DDT. Reproduction has increased since the mid-seventies. Now, within the park there are about thirteen pairs of Bald Eagles nesting, usually in tall lodgepole pines. One pair nests within eight miles of Old Faithful. For the eagles, the park is a marginal environment — the ice does not leave the lakes of this high, cold region until the end of May or early June. Turbulence in the rivers, the entry of streams into the lakes, and a few shallow sandy shores do, however, open up some water by mid-April, and this is important for the reproductive success of the eagles.

With protection following the 1940 Bald Eagle Act and the banning of DDT in the early seventies, the park eagles, even with marginal productivity, have served as the nucleus to enable a substantial increase in the number of breeding eagles in areas adjacent to the park: from about 16 pairs in 1960 to 36 pairs by 1982 (so that by 1982 the entire Yellowstone Ecosystem had 50 pairs).

The Boreal Forest of Interior North America

Michigan and Wisconsin
South of Lake Superior, the southernmost portion of the long arc of the Canadian Shield descends into the United States. Here in the lake and forest country of northern Michigan and Wisconsin, there are still many Bald Eagles.[17] For a time during the late sixties, it appeared as if these birds might go the way of those in other regions where there were severe effects of DDT. In the larger water bodies, DDT reached levels that essentially eliminated breeding eagles from the shores of lakes Superior, Michigan, and Huron, as documented in careful studies by Sergej Postupalsky, a tall East European immigrant who has become an internationally known raptor authority and has specialized in improving nesting opportunities for Bald Eagles and Ospreys in Michigan. However, inland from the large lakes, eagles survived near smaller water bodies where there was less buildup of DDT. In the years since 1972, the breeding success of Bald Eagles inland from the Great Lakes has improved on the Upper Peninsula of Michigan and in Wisconsin (by 1987 there was a total of 431 occupied nests in the two states), but it is probably still below that of the years before DDT. Breeding along the shores of the Great Lakes was the slowest to recover,

but even there a dramatic increase has occurred. In Michigan there were only 9 occupied nests along the Great Lakes as recently as 1982. In 1987 there were 30, and they produced 30 young. Nevertheless, high levels of PCBs and DDE in ten infertile eggs collected in 1986 from six Great Lakes nests suggest that enthusiasm for this increase should be tempered with caution.

Wisconsin has management plans for each nesting territory. Michigan intends to use a similar program, and through a computerized system and the authority of the State Endangered Species Act, the State Department of Natural Resources hopes to curtail threats to nesting success such as public access roads, clear-cutting, and snowmobiling projects.

Chippewa National Forest, Minnesota
In north central Minnesota, near the headwaters of the Mississippi River, a series of lakes on the southern edge of the boreal forest provide important breeding habitat for Bald Eagles (136 breeding sites in 1987).[18] The 89 nests where 151 young were raised in 1987 represent a considerable increase from the 20 to 30 successful nests found in the late sixties. Eagles generally arrive back on their breeding areas in mid to late March. Magnificent, towering white and red pines are the common nest trees; less often aspens are used. These trees are desired not only by eagles but also by the forest industry. Logging and extensive use of the waters for recreation led John Mathisen, a wildlife biologist with the forestry service, to develop effective plans for protecting eagle nests that have become the prototype for the rest of the continent. Under the program, all eagle nests on the Chippewa National Forest were identified and then a management plan was drawn up for each. The general guidelines employed a 100-meter zone around each nest where no activity was permitted, a 400-meter zone where no activity was allowed during the breeding season (February 15 to October 1), and additional protection up to 800 meters under some circumstances (Appendix 11).

Northwestern Ontario
Bald Eagles once nested around lakes and rivers across Ontario, from the Rideau lakes and the Thousand Islands region south of Ottawa all the way to the Manitoba border. Now, the birds are virtually gone from the eastern two thirds of the province, but they remain common in the northwest, particularly in the Lake of the Woods and Lac Seul districts.[19] Here in terrain typical of the Canadian Shield and its boreal forest, Jim Grier has banded and studied eagles for the past twenty years. This area was beginning to be affected by DDT in the late sixties. Fortunately, the reproduction of the birds has since recovered. Using random sampling aerial surveys of much of northwestern Ontario and part of adjacent Manitoba in 1974 and 1978, Grier has estimated that 500 to 1,000 pairs of Bald Eagles breed in the region.

Saskatchewan

Although most of the southern half of Saskatchewan is open prairie and farmland, the northern half, a land of lakes, forests, and rock, has many eagles, about 2,500 breeding pairs.[2] The Churchill River, really a series of lakes connected by rapids, flows across north central Saskatchewan within the boreal forest and, for the most part, within the rugged, rocky Canadian Shield. In the center of the province, one of the large lakes of this drainage system is Besnard Lake. Here, since 1968, we have studied Bald Eagles. As is typical of the lakes in the area, Besnard has an irregular shoreline and many islands. Even though its water area is only 62 square miles, the shoreline is 250 miles long. From 1973 to the present, the population of eagles on Besnard Lake has remained remarkably constant. In an average year 22 pairs occupy territories. Of these, eggs are usually laid in about 19 and young successfully hatched and raised in 16. Up to 29 young have been produced in a year, with 26 being the more usual number.[20] The lake also has many nonbreeding eagles. In July and August the population (excluding nestlings) reaches about 100 eagles. As in most other parts of these birds' range, nests in Saskatchewan are generally close to water (90 percent are within 220 yards of a lake or river), though occasionally eagles may nest up to a mile from a lake, particularly if there is a muskeg marsh between the lake and the nearest trees suitable for a nest.

The Pacific Coast

Washington, British Columbia, and Southeast Alaska
Along the coasts of British Columbia and Washington and their many offshore islands, Bald Eagles nest in large numbers (about 150 breeding pairs in Washington and about 4,000 in British Columbia).[21] On the San Juan Islands in Washington, all Bald Eagle nests found by Retfalvi were in Douglas-firs, usually in the largest tree in a local area, with an average height of 103 feet. Large numbers of Bald Eagles nest from the Alaskan panhandle and its border with British Columbia, all along the vast sweep of the southern Alaskan coast and peninsula, and westward almost to the end of the 1,000-mile-long Aleutian Islands chain. Estimates have ranged up to 50,000 birds.[22] Inland from the coast, eagles can be found along the major rivers, though the number of birds generally decreases progressively as one goes farther north. Fred Robards and John Hodges have flown thousands of miles in southeast Alaska, and by 1982 they had cataloged 3,850 eagle nests. For large areas of Admiralty Island and vicinity, nesting densities average as much as one nest per mile of shoreline, perhaps the densest population anywhere, with the possible exception of historical eagle numbers in Chesapeake Bay. Most of the nests are in large, old-growth Sitka spruce or western hemlock, often with a broken or bushy top and a large limb for support. Typically the nest tree is on an offshore island, particularly along broad channels between islands or on the side of an island

facing the mainland. Sometimes nests are on saltwater bays. Often a nest is on a prominent point or islet with an expansive view of the nearby water. Few nests face directly onto the open ocean.

It was southeast Alaska where much of the slaughter of eagles occurred between 1917 and 1952, the years of the Alaskan eagle bounty. The birds have now clearly recovered and are doing well. A decrease in eagle productivity in 1979 and 1980 (only about 20 percent of nests had young in the area) may indicate that these eagles have reproduced to the point that the population (more than 7,000 adults) is as large as can be supported by the existing prey base. There is, however, concern for the future because more and more forests are being clear-cut, thus removing potential nest trees. Fortunately, existing nest trees are now protected.

Amchitka Island, Alaska

The Alaska Peninsula and the Aleutian Islands extend more than 1,500 miles south and west out into the Pacific Ocean. In the last third of this long chain is a rugged, treeless island, so desolate that it was picked for a series of nuclear explosions that took place from 1964 to 1971. The tests — named Longshot, Milrow, and Cannikan — brought with them an army of people who came to study their impact on the environment; among them, Steve Sherrod, Clayton White, and Francis Williamson concentrated on Bald Eagles.[23]

Low, irregular cliffs make up much of the coast of the island. Narrow, ridged peninsulas project here and there into the ocean. Sometimes, a peninsula has been partly worn down. A pinnacle, or sea stack, connected only by a low saddle-shaped arm to the mainland, remains. Sometimes, erosion by the oceanic currents has left only a tiny, isolated islet. Such erosion takes many forms, and sometimes the base of the rock outcrop-

A nest with one chick on a grass-covered "peninsulet" on Amchitka Island, Alaska. The intertidal zone is visible beyond the nest (Clayton White).

ping has been eaten away far more than the top, leaving a flat-topped remnant that is ideal for a nest. It is here, on the sea stacks, ridges, and offshore islets, that most eagles nest. A few, taking advantage of the local elimination of the predatory arctic fox, also nest inland on hillsides.

Beneath the coastal cliffs, all around Amchitka, the daily ebb and flow of the tides alternately exposes then covers a wide intertidal region. Harbor seals, sea otters, Steller's sea lions, numerous gulls, and waterbirds frequent this zone, which is rich in nutrients and food. Beyond the steady line of swirling, tumbling waves is more eagle food—fish, ducks, and seabirds. The land is cool and desolate, but it is ideal for eagles. Their nests, one per 1.8 miles of shoreline, are closer together than on most other Aleutian Islands. They are usually within 20 feet of the shoreline and rarely more than 65 feet away. The proximity of the nests to the ocean may be an advantage when the eagles have to hunt on days that heavy fog crowds close above the sea. A thin fog-free layer of air just above the water may sometimes allow eagles to hunt, but one of the mysteries is how eagles navigate through fog, for they will fly even when the fog is thick, as it often is at Amchitka. Though the island itself is only 38 miles from end to end, with about 100 miles of shoreline, the intertidal region and the ocean just offshore are rich enough to support a population of 180 to 260 eagles.

6. A Matter of Space

Now pois'd and balanc'd in mid-space,
As resting from his airy chase.
Isaac McLellan

Gary—

The afternoon sun baked the lichen-covered rocks that I used as my eagle-watching perch. The heat was welcome, though, for these few weeks in July were the only period of the year that was reliably warm at Besnard Lake's high latitude. I was settled on a high rock outcrop about four hundred yards southeast along the shore from nest U. In a breathtaking view to the south, the main part of the lake extended into the distance. Below me a channel led to the bay where the nest was located. My job was to follow the comings and goings of the breeding adults, especially where and how they fished. Hidden inside a crude pile of sticks we called a blind, my assistant on the opposite shore kept a keen eye on the activities at the nest.

On a perch just above the nest, the female eagle, one leg tucked up under her belly feathers, dozed off and on. Her five-week-old chicks rested too, occasionally lifting a head and preening their newly sprouted feathers. I kept my eye on the male perched nearby on the uppermost tip of a tall white spruce, and from time to time I scanned the territory for other birds, eagles or otherwise, that might be of interest. It was a slow, lazy afternoon. The birds seemed content to perch. This was the curse of eagle watching—long hours of inactivity, enough to strain even the most avid field biologist. The rocks got harder (or my behind softer) and my attention span shorter as the hours passed. My mind drifted to thoughts of loved ones and Mum's home cooking; isolation was part of the price of studying eagles.

Suddenly, two eagles appeared directly in front of me, soaring. They were coming along the hillside from the southeast and going toward the nest. One, an adult, was at eye level, and the other, an all-brown immature, was fifty feet higher up. I cursed myself. Where had they come from? I quickly scanned around. The male's perch was empty.

Somehow something wasn't right. The adult was soaring lazily under the immature. Why wasn't he chasing the younger bird out of his territory?

Suddenly out of the sky fell a brown and white rock. At least that was what it looked like. It took a moment to grasp that it was nest U's male, head pointed earthward and wings tight against his body. He zoomed down — down, down, toward, at — into the adult intruder. At the last moment, the intruding adult rolled on its back and thrust its talons skyward. The resident adult somehow in a blur of motion pulled out of his stoop. A collision was barely avoided. In a tight loop he gained altitude. In doing so he passed the immature eagle and again paid no attention to it. The intruding adult meanwhile righted itself, side-slipped, banked sharply, and turned back in the direction from which it had come. The territorial male, however, was now literally on its tail in hot pursuit. Both birds flapped furiously. The pursuer gained altitude slightly. He dived again. The intruder, just far enough ahead, avoided a hit. Five or six times nest U's male attacked. Each time it seemed as if the intruder must be struck from the sky. Finally, almost out of binocular range, the territorial eagle relented and circled upward. With wings outstretched to their fullest, he soared higher and higher.

I strained my eyes to keep track of him, now only a brown speck against the white, billowy clouds. Pulling in his wings slightly so the tips pointed backward, he began to descend — a long, smooth glide back. A few minutes later, he landed on his favorite perch, gave his feathers a ruffle and shake, and settled down. If birds had emotions, this one had reason to be proud of himself.

My head still swimming with the excitement of the chase, I picked up my pencil and my notebook and sighed. Now I had to put it all down on paper.

*B*reeding Bald Eagles are territorial, that is, they defend an area around their nests from intrusion by other eagles. Although adults will, on occasion, vigorously chase and exclude other eagles, such an event is relatively uncommon. Most territorial defense is done in ways that require less effort. Adult Bald Eagles spend a good portion of their day perched at the top of prominent spruce trees not far from their nest. We believe that this serves as a signal to any would-be trespassers. The bright white head is as effective as any lighthouse beacon for marking a piece of land to be avoided; to another eagle the message must be quite clear — "Stay away!" An eagle flying toward a nest site can see the conspicuous resident and change course before it gets close enough to instigate a fight. Even we mere humans can spot the brilliant white of the adults from quite a distance.

A territorial adult calls excitedly to an intruder (Gary R. Bortolotti).

Territorial birds may also see intruders at great distances and warn them to keep out. Eagles often use a loud, ringing territorial call, or song, to declare their occupancy. A bird will toss its head skyward three or four times in rapid succession, emitting short, loud grunts followed by a high-pitched *whee-he-he-he* (which sounds something like the neigh of a horse). Breeding adults may also advertise their presence by soaring above the nest site.

In 1979 I conducted a few experiments to see how territorial eagles responded to tape recordings of their calls. At nest J, on a tiny island, I played a short segment of *whee-he-he* at normal volume. The old bird perched high atop a spruce immediately took notice. It swung its head around—left, right, left, right—looking for the intruder. I played the tape once more. Again it responded with agitated scanning. Then the bird took off flapping vigorously, flying in tight circles around its island, calling loudly the whole time. Around and around it went, obviously excited. I wondered whether the adult was more upset at the presence of another eagle in its territory or the frustration of not being able to see it.

Neither the white head nor vocalizations are sufficient to repel outsiders all the time. Occasionally, eagles resort to more physical means. A short flight from the nest may be all that is necessary. Sometimes interlopers casually soar across a territory and are silently escorted away by a member of the breeding pair. If you hadn't kept close track of who was who, you might think that the two birds soaring together were the residents. On other occasions territorial eagles launch a full-scale attack. The ultimate interaction between eagles is a spectacular display known as whirling, cartwheeling, or talon-grappling. An eagle in flight descends upon another airborne eagle. The lower bird rolls over and thrusts its talons upward. The two birds grasp each other's feet and then tumble earthward, separating before they hit the ground.[1] Many people believe this to be part of courtship (see Chapter 8), but it also, perhaps most commonly, appears in aggressive encounters.[2]

Observations of territorial behavior are sometimes puzzling. Occasionally, some outsiders appear to be tolerated closer to the nest than others, just as some are only scolded vocally and others are attacked. Observations of a pair of Besnard Lake's breeding adults that carried radiotransmitters suggest at least a partial explanation. The male was most likely to defend the territory against other males, and the female defended it against other females.[3] Sex-specific defense may be the best way of matching the size and strength of the resident to the opponent. Some eagles may be tolerated closer to the nest than others for lack of the appropriate territory holder in the vicinity to chase it away. Contrary to what one might think, it may not always be advantageous for the larger bird (i.e., the female) to be the aggressor. Although females probably have an advantage over males when it comes to strength, they are not as agile in flight and so may be at a disadvantage in aerial contests.

Various factors, such as whether the breeders are hungry or tired or whether the intruders try to catch a fish, may determine the nature and frequency of territorial aggression. Generally, defense is strongest against adults and weakest against the youngest immatures.[4] Adults with eggs or young usually continue active defense of their territory until after the young have left the nest, sometimes until they leave the lake.[5] However, when a pair has failed in its nesting attempt, the birds often continue to occupy

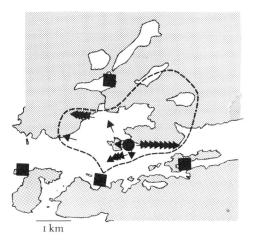

1 km

Approximate home range of an active breeding pair of Bald Eagles at Besnard Lake, based on fourteen days of continuous observation in July 1982. Nest shown by dark circle; active nests in neighboring territories shown by dark squares. The locations of aggressive interactions between the territorial adults and other eagles are shown by arrows (length of arrow shows the area covered during the dispute). In this case most of the home range was defended.

the general area of the territory, but no longer do they chase away other eagles.

Observations in various parts of the species' range suggest there are regional differences in territorial behavior. Some researchers have reported territories of nearly a square mile in size, based on observations of aggressive encounters that occurred up to about half a mile from the nest.[6] Territories, however, are probably not circular, except perhaps where nests are on small islands; pairs probably defend more water space than inland area. Eagles on Besnard Lake have territories of about two square miles in size. Eagles along the coast of British Columbia are said to defend an area shaped like a cone, extending above but not below or out from the nest.[7] Perhaps the structure of the habitat in this area (tall nest trees) has something to do with their seemingly atypical behavior. On Amchitka Island, Alaska, nesting adults pay little attention to other adults flying through territories, and immatures often perch only thirty to forty feet away from a nest.[8] Perhaps in dense breeding populations like that on Amchitka, it takes too much time and energy to try to exclude all (of many potential) intruders from a territory, and so pairs are more tolerant. In addition, if food is plentiful enough to support such a large population (see Chapter 5), then there may be less reason to be territorial. Presumably, part of the function of territorial behavior is to ensure sole access to a food supply.

Most territories are traditional, and their locations are remarkably sta-

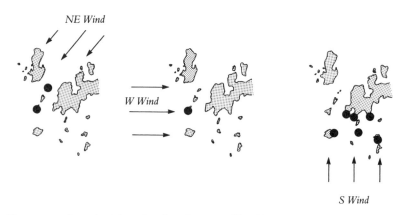

NE Wind

W Wind

S Wind

Home range of a territorial, nonbreeding four-year-old Bald Eagle (Cindy) on Besnard Lake. The perches she used when the wind was from various directions are shown by the dark circles.

ble over time. Even if a nest is destroyed, a new one is generally built nearby, and so roughly the same area of the lake is occupied by eagles year after year. It has always been assumed in eagle lore that because a pair of eagles breeds each year at the same nest, they are the same individuals. We now have evidence to suggest that this is true.[9] For some time we noticed that the temperament of birds, for example, shy or aggressive, was the same year after year at a particular territory. From 1980 to 1982 we quantified nest-defense behavior. Each time we climbed to a nest we gave the adults a score between 1 and 5; the higher the score, the more aggressive the reaction. Scores for each nest, averaged over ten or so visits per year, were remarkably consistent both within and among years. For example, a pair at one nest ranked 1.9, 1.8, and 2.0 over the three years, another ranked 3.6, 3.6, and 3.8, and yet another ranked 4.4, 4.9, and 4.8. More definite proof of nest-site occupancy by specific individuals has been the unique plumage markings of one adult and color markers on two females.

There are about 25 traditional territories on Besnard Lake. The breeding population has quite effectively carved up almost all the available space on the lake. There is little room for other pairs of eagles to become established. Occasionally, however, the space between pairs is such that new birds can squeeze in. In one case, it was a lone bird — Cindy, the color-marked four-year-old female described at the beginning of Chapters 3 and 4. She set up a territory among a group of islands right next to our base camp and bordering five breeding pairs.[10] Cindy reacted to the presence of other eagles up to 1.2 miles away — sometimes aggressively but sometimes amorously. She appeared to be attracting a mate, for courtship behaviors such as mutual stroking of the bills and copulation solicitation were observed (see Chapter 8). We had no idea that a lone female would defend an area. Was this typical behavior for a young adult? Little is known about Bald Eagles at that stage in their lives. Cindy left her territory on June 21 after at least a month's residence; why, we do not know, but a poor food supply may have had something to do with it. On June 20, we climbed four nearby nests and found no fresh prey remains in any of them. The nearby pairs may have been expanding their territories to feed their rapidly growing young and so squeezed out the single bird.

Occasionally, new territories are established. They may be in areas like the one Cindy occupied, just far enough away from active nests so that other adults would not find it worth their while to defend. In fact, the area where Cindy lived in 1978 later became home to a breeding pair. In 1980, two eagles were often heard calling to neighboring pairs, but there was no nest in the area. In 1981, a pair built a partial nest and occupied the site for the summer. By 1982 the nest was complete and contained a single egg. For some unknown reason the pair deserted their clutch; it was rather late in the hatching season and all the other nests had young. The fact that this was the only one-egg clutch I was to see in three years of climbing to nests, its failure late in the season, and the gradual nature with which the territory was established, all suggest that these birds were first-time breeders.[11] Successful reproduction followed in subsequent years. Data from watching color-marked birds also suggest that it may take time to establish a territory. The female at nest Y (described at the beginning of Chapter 8) is a good example. Hatched in 1973, she was sighted on Besnard in 1975, 1977, and in the vicinity of nest Y in 1978. In 1979 the marked bird courted a mate, and the two spent the summer occupying but not breeding in the Y territory. Finally, at seven years old, in 1980 she successfully reared two chicks.

Observations of the color-marked birds suggest that Besnard Lake eagles may not breed until six or more years old, that is, two or more years after attaining the white head and tail.[12] Therefore, the population as a whole is made up of breeding adults, nonbreeding adults, and nonbreeding immatures. That is typical of all healthy eagle populations in near-pristine environments. From our summer boat censuses of Besnard Lake, we found that immatures make up about 40 percent of the entire summer population, adults not associated with nests comprise 19 percent, with the remaining 41 percent being the breeding adults.[13] In southeastern Alaska, more than 50 percent of the adult population may be nonbreeders.[14] Such environments are saturated with birds. However, where lots of unoccupied habitat exists (for example, in areas where the eagle population has been greatly reduced by humans), younger birds, even immatures, may breed. Of seven eagles that were part of release programs (see Chapter 12) and have nested, five first bred when four years old, one at five years, and one may have bred when only three.[15]

Because so many adults in dense breeding populations do not breed but probably would if they could, territories are at a premium. Therefore, once an eagle is fortunate enough to get a territory, it must hold on to it. Generally, this means that pairs occupy the area they have claimed whether they are actively rearing young or not, probably to ensure that no other eagles move in and take over. Where possible, territorial adult Bald Eagles (particularly south of Canada and in mild climates) remain relatively near their nest all year long. Local conditions, especially food availability, likely determine whether territory holders move with the change in seasons. In the Greater Yellowstone Ecosystem, for example, eagles of a high-elevation breeding population leave their territories to spend the winter elsewhere, but those of a lower-elevation population do not.[16] The Bald Eagle's strong site tenacity can be a disadvantage; pairs are sometimes reluctant to relinquish their territory even when habitat changes have eliminated previously used and potential nest trees.[17]

Only part of the area where nesting eagles travel in the course of their daily activities may be defended from intrusion by other eagles (the proper definition of the term "territory"). The entire area, the "home range," may be larger than the defended space. Presumably there is a trade-off between the cost of defending an area and the benefit a bird acquires by having that area. Unlike territories, home ranges of adjacent pairs may overlap. That is not to say that one may find members of two pairs side by side on any given day. Where a nesting eagle may be found within its home range is strongly influenced by the direction of the wind. Eagles take the route of least resistance when they can and so soar along the shorelines where winds sweep across the open water to create updrafts. Consequently, home ranges are plastic, changing shape with the prevailing winds. Neighbors generally shift the same way with the wind and so are less likely to encounter one another in their travels.[10]

Factors determining how much space, either as a territory or a home range, each pair of eagles requires are not well known. The density of breeding Bald Eagles, a rough indicator of territory size, varies a great deal among lakes even within a relatively small area of Saskatchewan. Besnard Lake is particularly well endowed with nests, with one for every 11 miles of shoreline. Some birds nest less than a mile (as the eagle flies) away from another pair. The nesting density of eagles on lakes and rivers across the province varies from one breeding pair for every 10.5 miles of shoreline in the southwest to one every 71 miles in the northeast.[18]

One factor that appears to be important in determining the number of pairs that can "fit" into a lake is food. It seems that the more productive the fisheries, the more breeding pairs of Bald Eagles a lake can hold.[19] Many times I have talked with visitors to Besnard Lake and expressed my concern for the future of the eagles there. The popular fishing and recreational area is drawing more and more people each year, and I wonder how long the fish will remain plentiful. A common response is "Oh, well, the eagles will just move elsewhere if we overfish here. There are plenty of lakes in this part of the world." Unfortunately, many people fail to realize that

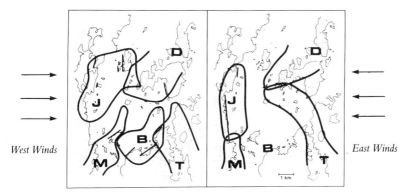

West Winds   East Winds

Approximate home ranges for five pairs of Bald Eagles raising young on Besnard Lake. The portion of the lake traveled by each pair and the shorelines used for soaring (shown by the solid bars) varied with the direction of the wind.

there is nowhere for these birds to go. All the other suitable lakes already have a resident eagle population that will not let new birds in. Each lake can support only so many eagles. Breeders displaced from one lake will probably join the ranks of adults that have no territory. As each lake dies, the eagle population is diminished.

At Besnard Lake, adults that do not have a territory may perhaps, like Cindy, take up temporary residence in parts of a lake where there are no active nests. However, most birds likely move around and exploit the best

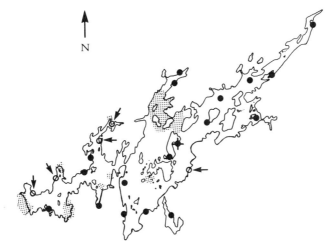

Distribution of breeding pairs of Bald Eagles at Besnard Lake in 1978. The dark circles show active nests, the dark circle with a + through it is an active nest that failed, and light circles show nests that were occupied by pairs but did not contain young. The stippled areas are portions of the lake frequented by immatures (i.e., generally where there were no active nests).

An immature Bald Eagle perched low and inconspicuously
(Gary R. Bortolotti).

food source available at the time. In spring and into early summer many of the adults not associated with nests, as well as the immatures, reside on small lakes adjacent to Besnard in proximity to spawning streams and away from active breeding territories. In July and August there is an influx of these nonbreeders onto the big lake as dead fish become more plentiful there (Appendix 4).[13] Certainly, the immatures seem to be true wanderers, never taking up residence in one area for more than a day or two. They drift from one lake to another and stay only while the food supply is good. Immatures, often in loose groups of two or three birds, are most often found in areas of a lake where there are no nests or in areas where there is a nest but the resident pair is not actively breeding.[13]

Unlike territorial adults, which perch conspicuously at the tops of spruces, immatures perch low in the middle of the crowns of trees along the shore. Their brown speckled plumage makes them much harder for us to spot than adults. Immatures may actually be hiding from adults; they certainly don't have to hide from predators or conceal themselves from wary prey. Young eagles are frequently victims of piracy by adults who have superior strength, speed, and agility. It has been our experience that in aerial contests between immatures and adults, the younger birds do not fare well.

The definition of territoriality sometimes includes the exclusion of other species from an area. Other raptors and particularly ravens are often at odds with eagles, but such birds are usually not chased out of the territory. I have seen White Pelicans and some gulls chased away, especially if a dead fish is floating near the nest (these birds are fish eaters and hence potential competitors with eagles). Another fisherman, the Osprey, has a long history of incompatibility with the Bald Eagle. Few Ospreys can pass up an opportunity to dive-bomb a perched eagle; however, eagles dominate the much smaller fish hawks and frequently rob them of food. Osprey nests are relatively rare in Saskatchewan. They can generally be found in the northern part of the province where roads have been built but are rare in wilderness areas suitable for eagles. In contrast, nesting eagles are absent from the human-developed areas but are abundant in remote locales.[20] It is possible that Ospreys have a tough time living on a wilderness lake because they are no match for the eagles there; instead, they establish themselves where eagles have been driven out by people. Ospreys are certainly much more tolerant of humans than Bald Eagles are. Alternatively, as humans move into an area they may alter the habitat, perhaps by removing the larger fish, and so make lakes more suitable to Ospreys and less suitable to eagles. After watching the two species interact (see Chapter 4), I can't help but favor the idea that eagles actively exclude Ospreys.[21] Consistent with this concept, the few remote lakes where Ospreys do nest are marginal habitat for eagles because of their late thaw (Ospreys nest later than eagles and so can breed successfully).

7. Eagle Architecture

High soars a patriarchal oak,
Its umbrage scath'd by lightning-stroke,
Upon whose topmost bough doth dwell
An eagle, monarch of the dell.
Isaac McLellan

Gary—

Our boat was full speed ahead on a collision course with the shore; I was
going to climb nest B and there was no time to waste. Before the boat
even came to a halt I leapt to the shore, leaving my assistant Ginny to tie
up. Stumbling over rocks and fallen logs I ran to the nest tree as if my
life depended on it. Every second counted in a race against time. If I was
to document the growth of eaglets without undue disturbance, the eggs
and chicks would have to be measured in as little time as possible. Spikes
driven into the big, old aspen afforded me a means to climb rapidly. I
almost ran up the first fifty feet. My heart raced as I neared the nest. I
knew as I looked up at the huge mass of sticks that it wasn't going to be
easy.

Just under the nest I paused to catch my breath. How was I going to
get over the rim? Year after year eagles had used this nest, and each time
they had added new material to it until now it was more than five feet
deep. I had to find a handhold I could trust or I would never be able to
work myself over the top. The trunk was so embedded in the nest that I
couldn't put one arm around the tree for security. In an effort to wiggle
my hands in around the bole I pulled at the branches and twigs. It was
futile. Most of the sticks were an inch or more in diameter and about
three feet long. Together with grass, moss, and other vegetation, the
sticks were woven and compressed into a solid mass, like an earthen
concrete. The grass and dirt I loosened blew into my eyes. With my arms
outstretched to the side, embracing the nest, and each hand grasping at
the thickest sticks of the nest, I held my breath and stepped higher onto
the terminal spike, my last secure foothold. I pressed my body close to
the nest and slowly snaked up higher, continually grasping new hand-

holds. The question whether I would be successful never entered my mind—I had to be. Finally, I worked up until I could reach above the rim to grasp the trunk as it emerged through the nest. Here, the tree was six inches in diameter. I held it securely with my right hand and sighed with relief. I tried swinging my left leg up over the rim, but it bounced off a stick jutting out from the edge. I tried again. No better. I was tired and scared. What now? As I pondered my position, the faint, constant sound of soft cheeps from inside the nest reminded me of the urgency of getting my work over with fast. With my left hand I broke the stick barrier and finally swung first a foot, then a knee, and then my whole body over the edge.

Any feelings of fatigue and fear vanished as my attention focused on the contents of the nest. In a shallow depression of an otherwise flat structure, two off-white eggs nestled in a bowl of dried grass. In one egg, a tiny bill protruded from a hole no bigger than the size of a dime. It uttered a soft peeping call. Above me both adults circled, cackling constantly and occasionally swooping down to within a few feet of my head. I had no time to waste. Quickly but carefully, I put each egg in a cloth sack and suspended it from a spring scale. I shouted the weights down to Ginny rather than taking the time to record them myself. I then took out a pair of calipers from my coat pocket and measured the length and breadth of both eggs.

My work completed, I left the eggs as I had found them, held on tight to the trunk, and swung my body over the edge. My feet searched for toeholds until I reached the first spike. Going down didn't seem as intimidating as going up. As soon as I hit the ground I ran to the boat. In one motion, I pushed off and jumped aboard as Ginny started the engine. I was absolutely exhausted, mentally as well as physically, and my stomach ached from the anxiety. However, the anguish of the climb seemed a small price to pay for what I had had the honor of witnessing. As we sped away, Ginny looked at her watch. Above the roar of the outboard she shouted, "Thirteen minutes from landing to takeoff—not bad!"

As a species the Bald Eagle appears to be an adaptable bird, even nesting on a giant cactus in Baja California, a bare rock in the middle of a Saskatchewan river rapids, and a hayloft of a barn on the Niagara River. However, when we look closely within a population, we see that the birds are highly selective as to where they place their nests. Bald Eagles generally build in a stand's dominant trees, often the survivors of previous fires and timber cutting; for example, big red and white pines in Minnesota, pines and cypresses in Florida, and Douglas-firs, Sitka spruces, and cottonwoods in Alaska.[1] An area's fire history may also help to explain an

A Bald Eagle's nest in the crotch of a trembling aspen on Besnard Lake. Note how both the area around the nest tree and the crown of the tree itself are open, thus allowing the parents an unobstructed path to the nest (Gary R. Bortolotti).

eagle's choice. In Saskatchewan, nests on the mainland are usually built in trembling aspens, and those on islands are in white spruces. Aspens are often the first trees to colonize a recently burned area, and much later the spruce will take over and dominate the stand. As trees on islands are less likely to be burned by fires that sweep across the mainland, forests on islands are usually older and are more likely to contain spruce than aspen.[2]

The actual tree chosen by eagles to bear their nest usually has distinct qualities. Whereas the Osprey typically builds at the very top of a dead stub or pole in an exposed location, the Bald Eagle prefers a more secure site in the crotch of a tree sheltered from the elements. An eagle's nest tree is usually alive but often has a broken, deformed, or dead top.[3] In southeast Alaska, nests in live trees with dead tops are used significantly less often than nests in trees with live tops.[4] Although eagles do not usually build a nest in a dead tree, they will continue to use an existing nest after the tree has died.[5] On Besnard Lake, extensive vegetation analyses by Anna Leighton and Doug Whitfield have revealed distinct differences between nest trees and trees selected at random. Nest trees, regardless of species, are stout for their height and have large crowns. A spacious canopy would certainly be advantageous for the adults during takeoffs and landings, as they have such a large wingspan, and a large crown would provide a secure site in which to build a large nest.

Only a few Bald Eagles have nested on man-made nest structures. Sergej Postupalsky has witnessed some nesting attempts on platforms in Michigan. Eagles have accepted a large tripod erected by Teryl Grubb in Arizona to serve as a substitute for a nest that blew down. Similarly, eagles have used makeshift wooden pallet platforms put up to replace fallen nests (see Chapter 12). A pair on Besnard Lake built a nest on top of one of my old treetop blinds even though there were two perfectly good old nests just two hundred yards away.

Across most of the Bald Eagle's range, it is common to find two or more nests within the territory of a single pair. Usually the active and alternate nests are within a few hundred yards of each other. Although multiple nests are characteristic of 38 percent to 53 percent of all territories in Washington and 33 percent to 80 percent of territories in the Greater Yellowstone Ecosystem, they are less common in Saskatchewan (only about 10 percent to 20 percent) and apparently do not exist for the eagles of Amchitka Island, Alaska.[6] It is curious that some pairs have two or three nests and others seem to be satisfied with one. Although it has been suggested that switching nests from year to year helps to reduce parasite loads, that does not appear to be an adequate explanation. An alternate nest may serve as a form of insurance — a place to go at the last minute should the pair be disturbed or if the primary nest falls down just before egg laying.[7]

Nest building is a part of the life of the Bald Eagle that is not often observed. As individual nests can be functional for several years, sometimes for decades, the building of a new nest is a relatively uncommon event. Francis H. Herrick, the Ohio ornithologist who pioneered the use of blinds to study birds, was fortunate enough to witness the building of an entire nest. One day, in the spring of 1927, a pair of eagles began to build a nest to replace their old one that had blown down ten days previously.

> On March 21 both birds were active in gathering sticks, mainly off the ground, and the following day was one of even more strenuous labor when one of the birds was often seen standing on the chosen site, receiving and placing sticks, or treading down the straw, which were brought in the talons of the mate; but all this action stopped when any one moved up to their tree for a closer view. After the third day their building impetus slackened, and by the end of the fourth the work of construction was virtually completed. The new nest appeared to be about five feet in diameter and four feet tall.[8]

In some parts of the Bald Eagle's range we are not sure at what time of year most new nests are built. Most of our work in Saskatchewan has been done from May to August, when the construction of new nests, at least complete ones, is rare (but does occur). Observations in September have

In the center of a nest, an egg and newly hatched chick lie in a grassy, cup-shaped depression (Gary R. Bortolotti).

suggested that many nests are refurbished and some new nests may be built then. Repairs to old nests and construction of new ones are also done in early April just before egg laying. Sometimes as much as one to two feet of nest material is added, although in most cases less.[9] Sticks comprise the bulk of the structure, but the interior of the nest is a mat of grass, moss, and sundry vegetation.

The purpose of building or repairing nests may not always be the immediate preparation of a site for egg laying. Pairs that are not breeding decorate their unused nest with a few green twigs during summer. It has been suggested that the activity helps maintain the pair bond. Other pairs may build "frustration" nests, entire or partial nests erected after a breeding attempt has failed, which may account for the alternate nests in some territories. Even immatures may be involved in nest building. At a few locations on Besnard Lake we have flushed two-year-old eagles from small, loose clusters of twigs in the crotches of trees. We do not know if the young eagles we saw had done the work or were merely attracted to the tree out of curiosity. One of the sites became a new and active nest the following year.

Nest material is brought to the nest by the breeding pair almost daily throughout much of the nestling period, and it accumulates to quite a depth by the end of the season. Fourteen inches of Spanish moss were added to one nest in Florida in just four weeks.[9] Some small twigs are carried in the bill, but most nest materials are carried in the feet. Occasionally, grass or twigs are taken from an alternate nest and brought to the active nest, within the same territory. Most vegetation is probably picked up off the ground, but some sticks may be broken off trees. I remember well how one adult female took off from her perch and in a burst of short, fast flaps flew right into the crown of a big, old aspen nearby. In midflight she grabbed a dead branch in her talons; the loud crack as it snapped could be heard right across the lake. With hardly a break in pace she flapped over to her nest and deposited the stick.

Sticks are generally placed around the perimeter of the nest, and other vegetation is deposited in the center. From my blinds, I got the distinct impression that many long branches were brought to the nest on windy days. My first thought was that the adults brought heavier loads under such conditions because the greater lift would be an advantage. However, the loads weren't really as heavy as they were bulky and awkward—the branches often had several twigs along their length. I watched the adults use them to form a high corral along the perimeter of the nest, and I began to wonder if this could be an adaptation to prevent the young from falling. Improbable, perhaps, but possible. For quite some time, the developing chicks are clumsy and could easily tumble over the edge, given a little help from a gust of wind.

Branches brought in summer are often from living trees, either aspen or spruce, in contrast to the dead ones used to build the bulk of the nest. There may be a special reason for this. Bringing greenery could function to maintain the pair bond or to inhibit parasites of the nestlings.[10] One clear result of delivering nest material throughout the breeding season is that the nest is a cleaner place to live. Old fish, rotting and infested with maggots, accumulate in the nest. The only form of nest sanitation appears to be burial of the waste under large quantities of grass, moss, and other greenery.[11] Perhaps eagles do not simply toss old fish over the edge because that might attract mammalian scavengers that could be predators of the eaglets. For example, on Besnard Lake many nest trees have been climbed by black bears.[12] In other areas, raccoons are suspected of preying on the eggs or young.[13]

When the adults are at the nest and not brooding or feeding the young, much of their time may be spent in arranging sticks and digging in the nest material. This digging is often done with great vigor. An eagle pushes its bill all the way into the grassy mat, then pulls out and shakes a huge mouthful of vegetation. This may go on for several minutes and be repeated several times a day. Consequently, the adults accidentally, or perhaps even deliberately, ingest some of the nest material, which is later cast up in a pellet.[14] The digging may further sanitize the nest, perhaps inhibiting the blowfly larvae that parasitize the eaglets, or it may dry out the grasses and improve the nest's insulating properties.[15]

An eight-week-old eaglet atop a pile of sticks on the ground that was, until a few days previously, its nest seventy feet up in a trembling aspen. The nest tree was toppled in a windstorm. The eaglet was not injured and made its first flight two weeks later (Gary R. Bortolotti).

One of the consequences of annual and seasonal accumulations of plant material in traditional nests is that the structures can get very big. In fact, the world's largest bird's nest was built by a Bald Eagle at St. Petersburg, Florida; it measured 9½ feet across and 20 feet deep and was estimated to weigh more than two tons.[16] Another remarkable Florida nest, built 115 feet above the ground in a cypress, was 10 feet wide by 12 feet deep.[9] The Great Nest in a dead shellbark hickory at Vermilion, Ohio, measured 8½ feet across at the top and 12 feet deep; it was occupied for 35 years before it blew down in a March storm in 1925.[8] Such masterpieces of eagle architecture are more likely to appear where eagles have access to large hardwoods, such as hickories, sycamores, and elms, or to big pines. Over much of the continent few of the big trees remain today. How this has affected eagles is difficult to evaluate. Some birds have certainly made do with lesser trees, but many forested areas have simply disappeared. On the Pacific coast, eagle nests still sway 200 feet above the ground in 400-year-old spruces and Douglas-firs. Such trees, much in demand by lumber companies, are not easily replaced.[17]

A typical nest in the boreal forest is built 25 to 70 feet up a trembling

A Bald Eagle's nest with two eaglets on a cliff in Arizona (Teryl G. Grubb).

aspen and is 5 feet wide by 3 feet deep. Aspens are relatively short-lived trees (although some reach 150 years of age), so there is little opportunity to accumulate many years of nest material. Of 48 nests on Besnard Lake for which we knew the date of construction, only half were still extant after 6 years.[18] Although support branches sometimes collapse under the weight of the nest, more often the entire nest tree is the victim of a windstorm. If the chicks are well feathered, they may survive the fall and continue to be cared for by their parents on the ground.

When the eaglets are large, it is safer (for them and the researcher) to lower them in a bag to the ground for banding and measuring. By late summer some nests, such as this one, begin to fall out of the tree (Gary R. Bortolotti).

Although cliff nests are more secure than those in trees, they have their own problems. In Arizona, where cliffs are actually preferred to trees as nest sites, the fledging success for pairs using cliffs is only about half that of tree nesters. Heat stress on the young is considered to be a major problem.[19] Nests on the ground are most commonly found where there is an absence of suitable trees and of mammalian predators (e.g., foxes), as is the case on some remote islands in the Aleutians.[20]

Even if a nest tree is secure, the nest itself may not be. Some Besnard Lake nests, particularly in their first year, are barely big enough for the parents and are crowded indeed by the time two eaglets are exercising their wings in anticipation of their first flight. Such nests can be so inconspicuous that each spring as we search for new nests, we sometimes spot the white head of the incubating bird before we see the bundle of sticks. By the end of summer many nests are dilapidated. Sometimes the nest comes completely apart, leaving the parents with the awkward chore of delivering fish to their young, which are old enough to perch on limbs but not old enough to leave the nest tree. We spent sixteen hours watching what happened after one nest fell out of the tree two weeks before the chicks fledged. The parents had great difficulty transferring food to their hungry offspring—six of seven fish brought to the nest tree were accidentally dropped to the ground and never recovered.

8. A New Generation

Skirting the river road, (my forenoon walk, my rest,)
Skyward in air a sudden muffled sound, the dalliance of the
    eagles,
The rushing amorous contact high in space together,
The clinching interlocking claws, a living, fierce, gyrating
    wheel,
Four beating wings, two beaks, a swirling mass tight grappling,
In tumbling turning clustering loops, straight downward falling,
Till o'er the river pois'd, the twain yet one, a moment's lull,
A motionless still balance in the air, then parting, talons loosing,
Upward again on slow-firm pinions slanting, their separate
    diverse flight,
She hers, he his, pursuing.
       Walt Whitman, "The Dalliance of the Eagles"

Gary —

I settled down on a rotting log and made myself as comfortable and
inconspicuous as possible among the alders that lined the shore opposite
nest Y. It was mid-May 1980 and my first sweep that year of the who's
who in the breeding population of Besnard Lake. I was particularly
excited about this nest. Last spring a female turned up on the territory,
and although she didn't raise young, she stayed the summer and courted
a male. She was special. On her left wing she carried a red vinyl tag that
Jon and Doug had given her as a nestling in 1973. There had originally
been three markers, but by 1979 only one remained; its edges were worn
and curled, and like an old tractor in the sun, the bright red had faded to
a dull orange. If the marker survived to this year, the bird would be a
valuable subject for my research. An opportunity to study a wild eagle
of known age and origin could not be missed.

I focused my telescope on the adult lying low in the nest. When I
landed the boat I had a good look at its mate, now off fishing somewhere, and there was definitely no marker on it. Except for its bright
white head the bird on the nest was barely visible. Its breast was sunk
low and its back lay flat, sure signs that there were eggs in the nest. If I

was going to see a marker, the bird would have to move. I anticipated no more than about an hour's wait. Incubating eagles generally stand up about once an hour to turn the eggs and poke around in the nest material.

The hours dragged on and nothing happened. For lack of something better to do, I took more and more detailed notes—"eagle looks to the NW," "eagle looks to the NE," "eagle yawns." Fortunately, the sights and sounds of the boreal forest were always entertaining. Oblivious to my presence, a snowshoe hare lunched on plants not an arm's length away. In turn, I was scrutinized by a clan of Gray Jays; I wondered if I amused them as much as they amused me. I even added a bird to my life list, a Harris' Sparrow passing through to more northerly breeding grounds.

In the bay that separated eagle from eagle watcher, three male goldeneyes courted two females. My alder cover afforded a delightful view of the spectacle of splashes, squeaky quacking, and contortions that made up the display repertoire of the drakes. Suddenly the scene exploded into a flurry of wings. A flash of brown swooped down on the ducks. I froze. It was an immature goshawk. Missing a goldeneye by a mere few inches, it flew to perch on a spruce on the shore not thirty feet from me. With eyes glued to its quarry, the raptor was far too intent on hunting to notice me. In less than a minute it launched another attack. With stiff, rapid wingbeats the goshawk shot out from its perch and headed down to the water at a steep angle. The ducks scurried and splashed about but did not fly. The attack was short-lived. The hawk pulled up and away long before it was within striking range. I suspect that if a goldeneye had taken to the air it might well have fallen prey to the accipiter. The goshawk reperched farther away now, and odd as it seemed, the drakes resumed courting. Perhaps predator and prey knew each other's abilities well enough.

The hawk did not dawdle or waste effort on the unattainable. Flapping across the bay, it landed sixty feet downshore from Y at about the same height as the nest. The eagle raised its hackles and just glared at so bold a bird. The two eyed each other for a minute and a half, then my jaw dropped—the goshawk flew straight at the incubating eagle. The eagle ducked as the hawk skimmed overhead. Keeping tight on the eggs, the eagle slowly raised its head, eyes locked on the attacker, who reperched just fifty feet away. Unfortunately for me, the excitement was to end there. The goshawk lost interest in the eagle and in a few minutes drifted along the shore and out of sight. Not even the goldeneyes seemed to notice.

Despite the change in pace, it became increasingly difficult for me to be patient. After five and a half hours, the bird still had not moved off its eggs. This was highly unusual. Incubating eagles usually stand up and poke around in the nest about once an hour. Eagles do sit tight when disturbed, but I believe I was well concealed and that my presence was an unlikely explanation for the bird's odd behavior. As eager as I was to see if the precious color-marked bird was breeding, I had to leave, for I had a dozen other nests to watch.

The next day I returned. The bird did indeed have a red wing tag. She was the same eagle seen in 1979, and she has since then nested in the Y territory every year up to the time of writing, in 1987.

*C*areful observations of the courtship behavior of wild eagles are few. It is a difficult subject to study for many reasons. In Saskatchewan, for example, Bald Eagles return from their wintering grounds in late March and early April. It is hardly most people's idea of spring: temperatures are below zero and the lakes still have a thick sheet of ice. The birds seem particularly wary of people, and as this may be a time when they are sensitive to disturbance, it is best to leave them alone. When observations can be made, they are not always easy to interpret. High-altitude soaring appears to be typical spring behavior, but whether it serves some territorial or courtship function or is merely the product of peculiar weather conditions is unknown. Males also make undulating flights, and pairs are frequently seen perched together. The whirling display, so beautifully described by Walt Whitman at the beginning of this chapter, is often mentioned as being part of courtship.[1] However, as whirling may also be seen when a territorial eagle attacks an intruder (see Chapter 6), there is much to learn about the behavior before we can say anything definitive about it.

Observations of a pair of captive Bald Eagles at the Patuxent Wildlife Research Center at Laurel, Maryland, have provided valuable insight into courtship activities that are difficult to see in the wild.[2] The male and female frequently perched side by side and stroked and pecked at each other's bills. The female used her bill to stroke her mate's back and breast, while the male stroked her head, neck, and shoulder. The Bald Eagle's bill may be so large and conspicuously colored to enhance such courtship displays. The preliminaries leading to copulation are described by Naomi Gerrard:

> The female approached first 8 of 16 times. She lowered her head, spread her wings slightly, and moved toward the male, opening and closing her beak appearing to call. Usually this calling was inaudible from the blind. When heard, it was a single, soft, high-pitched note repeated several times, very different from calls used on other occasions. She would then move her head near his wing and nudge him. When the male made the approach he hopped near her and then edged closer, sometimes calling in a similar fashion. Following the initial approach by the female, the male called, then flapped his wings and moved his tail up and down (usually lowering his body and hitting the perch with his tail). When the initial approach was made by the

A pair of Bald Eagles breeding at the Patuxent Wildlife Research Center, Laurel, Maryland (Naomi Gerrard).

male, the female positioned herself with feet further apart, head lowered, and wings slightly apart (bowing), ready for the male to mount her.[2]

On ten of the sixteen occasions that Naomi observed these behaviors, the preliminaries ended there and there was no copulation. Although males may not always be motivated to mate, sometimes the females' large size intimidates them into not performing (female raptors have been known to kill males in captivity). Even at the best of times copulation appears rather awkward. The male curls up his toes and steps onto his mate's back. Her tail goes up, his tail goes down, and semen is transferred when their cloacae (the vent for both reproduction and excretion) meet.

We have seen a few unmated females attracting a mate as well as soliciting copulation in the middle of summer. The color-marked bird at nest Y mentioned above was one, and Cindy, described at the beginning of Chapters 3 and 4, was another.[3] Pairs rearing young aren't the least bit interested in sex. As sexual behavior late in the season is inappropriately timed, copulation or copulation-like behaviors may have other functions related to the pair bond.

It is assumed that pairing is a long-term relationship (i.e., over several years). One may often see references to the "fact" that eagles mate for life, but no one really knows if that is true. (There is evidence that some birds occupy the same territory year after year; see Chapter 6.) There are, however, observations that suggest that the sex life of the Bald Eagle may be more interesting than previously suspected. Although the vast majority of Bald Eagle breeding attempts are probably monogamous matings, there have been a few cases, in the Aleutian Islands and in Minnesota, of three adults attending a single nest.[4] It is also probable, given recent studies on other species, that eagles "divorce." We found that generally in the year following a failed breeding attempt two eagles would occupy a territory but not try to rear young.[5] It is possible that after a failure the pair splits up, and so in the next year they are looking for new mates. Such a strategy makes sense; it might be better to start over with a new partner than risk another failure with the old one.[6]

In the case of divorce, a year of nonbreeding may be required to allow enough time to find a new mate and secure a pair bond. Under other conditions, it seems that pair formation can be quite rapid. If one member of a pair dies during an active breeding attempt, it may be replaced in a short time, but the stability of the new pair and the quality of the new mate as a parent are not known. It is also possible that some eagles may find mates in winter. Al Harmata saw pairs engaged in courtship in winter in Colorado. The stability of such relationships is uncertain, especially in light of the observation that males departed on migration alone and a few days ahead of the females.[7] Most true courtship is probably done on the breeding grounds just before the eggs are laid.

The timing of egg laying seems to depend a great deal on local conditions, probably meteorological, for birds in different geographic areas start nesting at different times of the year. Southern Bald Eagles lay earlier and have a longer laying season than northern birds.[8] For example, eagles in Florida usually lay from the first week in November to mid-December, but the range may be as great as five months.[9] In contrast, the females of 90 percent of the pairs on Besnard Lake lay within a period of about ten days beginning in mid-April.[10] The laying period may be so narrow because of the shortness of the summers in this northern locale. Laying seems to be timed so that the eggs hatch close to when the lake ice thaws. This ensures that the young will have as much time as possible to develop hunting and flying skills before migrating. Hatching before thaw may be risky, for open water is usually required for the parents to find an adequate supply of food for their young. The length of the breeding season may, in fact, limit the northern distribution of eagles. In Saskatchewan, the density of breeding pairs is correlated with the mean April temperature.[11] Lakes in the northeast corner of the province thaw a month later than those in the southwest, making them much less usable to nesting eagles.

The beginning of the laying period shifts slightly from year to year, probably because of the weather. At Besnard Lake, the relative timing of laying is remarkably consistent within a territory from year to year; that is, some nests are always early and others are always late.[12] This is no doubt because the same individual birds seem to occupy the same territory from one year to the next. Although we know this to be true for a few nests because of color-marked birds, it is also suggested by data on egg size. There is a remarkable consistency from year to year in the size of eggs found in a nest.[10]

Just as body size in Bald Eagles increases from the southern part of the species' range to the northern part, so does egg size. Eggs in Florida vary in size (length by breadth) from 79 by 56 millimeters to 58 by 47 mm, and those in Alaska and northern Canada varied from 84 by 60 mm to 70 by 53.[13] Of 27 eggs from Besnard Lake, the largest was 80 by 56 mm and the smallest 67 by 54 mm. The second egg laid within a clutch is slightly smaller than the first, and the third is smaller than the second.[10] At first glance, Bald Eagle eggs seem rather small for such a large bird. However, as a rule among species of birds, as body size increases the relative size of eggs decreases. An eagle's egg weighs barely more than 2 percent of the female's body weight, whereas a songbird's egg is typically about 13 percent, and several eggs constitute a full clutch.[14]

Factors that potentially influence the clutch size of birds have long been of interest to ornithologists. Bald Eagles generally lay two or three eggs. At Besnard Lake between 1980 and 1982, 8 nests contained three eggs, 25 had two eggs, and only 1 had a single egg (see Appendix 5). Three-egg clutches are concentrated in the early part of the nesting season. Similarly, Broley found that Florida eagles nesting early raised more offspring than those nesting later in the season.[15] Both Broley's work and our own suggest that some pairs consistently produce three-egg clutches. That could be a factor of the age or genetic makeup of the female or perhaps the quality of the spring or winter habitat.

When eagles return to frozen Besnard Lake, food is in short supply; however, some pairs have access to open water at small streams where spawning fish are abundant and easily captured. Although nest sites may not be chosen close to such streams, eagles breeding near them produce more offspring than more distant nesters.[16] It is not clear exactly in what way food may be important. For example, the nature of the resources available in the spring may potentially determine the number of eggs laid, hatching success (perhaps hungry parents are poor incubators), or maybe even whether a pair breeds at all.

The physical condition that females achieve on their wintering grounds may be just as important to reproduction as food supply in the spring. Eagles do not appear to feed much during spring migration. Instead, they may put on weight before departure.[17] It may be that only those birds that arrive on the breeding grounds with abundant energy reserves can nest successfully. Adverse weather conditions during spring migration could potentially affect reproduction by depleting those reserves. That was suggested to us by an anomaly in our data from Saskatchewan. Although the productivity of the Besnard Lake eagles has been remarkably stable since

A few miles from the space shuttle launch pad at the Kennedy Space Center, a female eagle incubates eggs while the male brings in a branch to build up the perimeter of the nest (NASA).

it was first documented in 1968, the 1975 breeding season was an exception. That year only 63 percent of the pairs had young, and only about half as many fledglings were produced as were normally expected.[18] Blizzards had raged across the prairies at the time eagles were moving northward, likely affecting both food availability and the physical condition of the females just before egg laying. Poor weather conditions in the spring of 1975 appear to have been widespread, and thus it is no coincidence that reproduction in the Greater Yellowstone Ecosystem was also disastrously low that year.[19]

Researchers have found that the ups and downs in annual productivity of Bald Eagle populations are consistent across several geographically disparate breeding grounds.[20] Their findings suggest that some widespread factors, such as weather, may affect Bald Eagle productivity over large areas. Whether spring or winter food is more important and what kind of role weather plays are not just questions of academic interest. If we are to help bring back the Bald Eagle and maintain healthy populations, we must first know where eagle reproduction is most vulnerable and then decide where our precious management dollars can be best spent.

Once eggs are laid, a pair of Bald Eagles settles down to a relatively quiet existence—at least we assume so. Little is actually known about egg laying and incubation in comparison with the nestling stage of reproduction. For all but about 2 percent of the day, one adult, either the male or the female, can be found diligently sitting on the eggs.[21] About once per hour, the incubating bird stands up. Sometimes it will change position or poke around in the nest material, but often it turns the eggs. Not only does turning allow for more even warming but it also prevents an embryo's membranes from sticking to the shell.[22]

For the few minutes of the day that the clutch is left unattended, the adults may cover the eggs with dried grass before departure.[23] The amount of time that the eggs were left unattended by the captive breeders that Naomi Gerrard observed was directly related to the windchill factor—the colder and windier the day was, the less likely the parents were to leave their eggs exposed (Appendix 6). In a Saskatchewan spring, incubating adults may have to endure snowstorms and other weather hazardous to developing embryos. When one adult relieves the other, both birds show the utmost care in stepping around the eggs. The feet are clenched so that the large hind claw is shielded by the front three toes.[24] The bird about to take over incubation slowly steps to the eggs, straddles the clutch, lowers its body, and finally settles down with a subtle side-to-side rocking motion. Unlike a brooding adult, which raises its breast slightly as it covers the tiny chicks, an eagle on eggs lies flat and low in the nest. Typically, all that is visible when viewing from the ground or a boat is the incubating bird's white head and tail.

While one member of a pair incubates, the other searches for food. Generally, each member of a pair of Bald Eagles feeds itself, unlike many species of raptors in which the female does not hunt at all during incubation but rather is fed by the male.[14]

The off-white eggs hatch after about 35 days of incubation.[25] However, hatching failures are common. About 8 percent of eggs laid at Besnard Lake from 1980 to 1982 were addled or rotten. They remained visible in the nest and may have been incubated for some time, even though the adults were feeding young as well (never more than one egg per clutch was addled). Infertility, perhaps from incomplete or improperly timed copulations, may account for the failure of some eggs to hatch.[26]

Human disturbance may also cause hatching failures, but it is more likely to affect the entire clutch rather than just part of it. Incubating eagles are particularly sensitive to the presence of people near their nests. A pair may desert their nest in response to a disturbance, or they may just be absent from their clutch long enough for the eggs to die from exposure. In Saskatchewan, snowmobilers, early in the incubation period, and boaters, late in incubation, are sources of problems. Each year on Besnard Lake, the eagles are exposed to a flood of disturbance on the opening day of the fishing season, a time when many eggs are hatching. We have seen some

A tiny bill has poked a hole in the egg. The chick just hatched the day before (Gary R. Bortolotti).

failures coincide with the large influx of people. Researchers too have inadvertently caused nesting failures by trying to count eggs from low-flying helicopters or by spending too much time at a nest. We found that the key to avoiding the deleterious effects of disturbance was to keep our visits (late in the incubation period) to no more than fifteen minutes, less if possible.[27] After the disturbance it was important that we go far away from the nest, often more than half a mile, so the adults would settle down as soon as possible. The length of time one disturbs the birds is critical when nests contain eggs or chicks less than about three weeks old. Although disturbance is never completely harmless, the chance of a serious detrimental effect on large nestlings is substantially less.

In the fifties and sixties, DDT and other contaminants were responsible for a much higher percentage of hatching failures than would naturally be expected. The political battle to ban DDT was helped immensely by evidence that eggshells were much thicker before DDT was used (before 1946) than they were afterward.[28] Proof was possible only because historically many clutches had been taken from the wild by people who collected eggs as a hobby, just as one would collect stamps. Over time, many eggs in private hands became part of the research collections of museums. It is ironic that the people who plundered so many nests played a key role, albeit indirectly and unintentionally, in saving the species.

There is yet another cause of hatching failure: egg burial. We have observed two interesting cases at Besnard Lake. One was at nest J in 1980. On my first visit to the nest I found three eggs, two of which were pipping, lying naturally in the nest cup. However, on my next visit, four days later, one egg was almost completely buried in the grassy nest material and was badly cracked. Inside that egg was a fully formed embryo of hatching age. The second incident involved nest T. On my first climb to the nest I was surprised to see only one newly hatched chick. I had always found at least two young or one young and an egg in other nests. No eggs were visible in this nest, nor did I find any after digging around in the upper three to four inches of nest material. One week later the nest failed; a nearby aspen crashed into the nest tree during a storm, and so the adults probably deserted or the chick died. I climbed into the empty nest and dug around the grassy nest material out of curiosity. I was flabbergasted to find an egg buried about six to eight inches deep in the center of the nest. I was even more surprised when I went to blow out the contents of the egg to save its shell for a museum. Inside T's eggs was a perfectly formed embryo that must have been ready to hatch.[29]

The eggs in nests J and T must have been buried by the parents just after the other eggs hatched. But why? Most likely, the adults were stressed for food. They "decided" that they could not raise as many offspring as they had eggs, and so when the first egg or eggs hatched the other was covered up. The parents did, in fact, seem to have a particularly difficult time finding prey. There were no food remains on my first visit to nest J. Similarly, the only remains I found in T were two old, bony fish heads. That was unusual. At eleven other nests we climbed to at the same time, there were between two and sixteen pieces of fish, totaling on average 2,855 grams (more than six pounds) in each nest. Whether food stress can cause Bald Eagles to abandon part of their clutch may never really be known, but it is an interesting possibility.

9. Growing Up

O'erlooking from his eyrie grand
The wide expanse of forest land.
Isaac McLellan

Gary —

Nest Delta: 3 June 1982. It was high noon and I had already been in the treetop blind for several hours. Things were quiet now. The sun was shining and the trees swayed gently in the breeze; it was as peaceful and serene a day as I had ever spent watching eagles. The adult female dozed off and on as she stood to one side of the nest. Her four- and three-day-old chicks huddled together in the grassy nest cup. Much later I determined that the older eaglet (which I routinely labeled C1) was a female and the younger (C2) was a male, but at this size — a mere 232 grams and 144 grams, respectively — they were too small to tell.

These little beige puffs of down were comical to watch. Their body proportions were not at all like a mature bird; a huge head and long legs were mounted on a pudgy little body. Soon C1 and then C2 squatted (they were much too young to stand); they held their heads upright while their legs lay flat in the nest in front of them. Facing each other, they stretched their necks vertically as far as they could go. I wrote down that the chicks changed posture and the hour, minute, and second that they moved. I waited for what would happen next. This neck-stretching behavior often preceded bouts of aggression. Usually the first-hatched eaglet bit or pecked the second-hatched eaglet until the younger bird lay submissively in the nest; however, on this day I was taken by surprise.

The heads of the eaglets swayed back and forth as if blowing in the wind. In a flash, little C2 lunged forward and bit C1 on the side of the head. The force of the attack knocked C1 backward into the nest. C2 held on. I could not believe my eyes; C2 was twisting and shaking C1 from side to side. This was no ordinary fight. C1 was trapped against the rim of the nest cup where she had fallen and now could not right herself. C2 took advantage of the situation and pummeled C1 — peck, bite, shake — over and over. It wasn't long before C1 began to call out with a high-pitched chittering sound. I had only rarely heard this distress call before. Peck, bite, shake; C2 showed no sign of diminishing his assault. To my surprise, the adult stepped over to her brood, bent down, and took C2's bill in hers. Gently she rocked her chick back and forth several times. C2's entire body moved from his mother's efforts, but it wasn't enough to make him relinquish his prize. C1 chittered even louder now as she flailed helplessly about. When would it all end? It seemed to go on forever. The adult gave up on trying to pull the combatants apart. Instead, she straddled her chicks and plunked down on top of them. Both chicks disappeared as the adult rocked from side to side in an exaggerated manner. Seventeen seconds later, or more than a minute after the launch of the attack, C2's head popped up through his mother's breast feathers.

I was bursting with curiosity. How was C1? Was she seriously injured? It wasn't until more than an hour later, when the adult male landed on the nest, that I finally got a glimpse of C1. As the brooding female stood, C1 squatted up, apparently no worse for the beating she had taken earlier. I had never seen anything like today's fight, and I never did again. C2 was absolutely subordinate to his big sister for the remainder of the time they spent together in the nest.

For eggs that are fertile and survive incubation, hatching is a long process. Twenty-four or more hours may pass from the time the eggshell is first pierced until the eaglet finally wriggles free. Wet and helpless, eyes closed but soon to open, the chick gives a soft peeping call almost constantly throughout hatching. As is typical of birds, breaking of the shell is facilitated by a small egg tooth. This bony little projection on top of the upper mandible is usually retained on eaglets for a month or more before it wears or falls off. The sight of an eight-pound eagle with an egg tooth is odd indeed.

Upon hatching, an eaglet is covered with a pale beige-gray down, the first of two distinct coats. The skin and scales of the legs are a bright pink, the bill is gray black with a white tip, and the talons are flesh-colored. By the end of the first week of life the skin is tinged with blue, the legs begin to yellow, and the bill darkens. Ultimately, over the next few weeks, the bill will become a dark gray-black, the legs bright yellow, and the talons black.

Because a female lays her eggs a few days apart and begins incubation with the first egg, hatching within a clutch is asynchronous. One or two days is the usual age difference between siblings, but I have twice observed eggs hatching on the same day. The age difference between siblings is one of the most profound determinants of the well-being and survival of eaglets

An eight-day-old eaglet squats in the nest. It is covered with its first of two coats of down (Gary R. Bortolotti).

A twenty-day-old eaglet. The first down remains in any quantity only on the top of the head. The bulge in the throat is the bird's crop, about one-half to two-thirds full here (Gary R. Bortolotti).

within a brood. Older chicks are larger and hence have a substantial advantage in obtaining food provided by the parents.[1] The longer the hatching interval, the more capable the older bird is of physically dominating younger nest mates. Death of the younger chick in two-chick broods is relatively uncommon, at least at Besnard Lake. However, the size disparities between youngest and oldest in a three-chick brood are such that the third-hatched chick has little chance of survival. For example, one nest on Besnard Lake contained chicks aged nine, eight, and six days. Their weights were 477, 260, and 80 grams, respectively. In other words, the youngest was about the weight of an American Robin, whereas the older was as heavy as a Broad-winged Hawk. That particular third-hatched chick had not gained any weight from the time it hatched; it was dead before our next visit four days later. If chicks are going to die from sibling competition, they almost always do so in the first week or two after hatching.

A brood of three nestlings. Note the large difference in degree of feather development among these siblings (Jon M. Gerrard).

From 1980 to 1982, we saw six nests in which three young hatched, and in only one of them did three young fledge. It may be that large broods can be reared only in years of exceptional food abundance. Over a fifteen-year period at Besnard Lake only eight nests, about 4 percent of the total number of nesting attempts (see Appendix 5), produced three fledglings; however, three nests (13 percent) did so in 1978 alone.[2]

Unlike the young of some species (such as the Black and Crowned Eagles of Africa, the Lesser Spotted Eagle of Europe, and sometimes the cosmopolitan Golden Eagle), Bald Eagle chicks that I saw die from sibling competition did not go out in a bloodbath. The dying young I've examined were never severely injured; instead they gradually starved. Fighting among chicks creates a dominance hierarchy, a pecking order. The older eaglets peck their siblings into submission and so prevent the younger birds from obtaining an adequate amount of food for survival. As sad as it sounds, remember that it may be a mechanism that adjusts the number of young in the brood to the food supply. Food shortages later in the season might cause the entire brood to perish if one chick is not eliminated early in life.

The fighting incident related at the beginning of this chapter was unusual. Although the second-hatched chick sometimes initiates a fight, it only rarely wins a battle. The incident was also peculiar in the direct manner in which the adult intervened. Perhaps because most conflicts last only a few seconds, it is unnecessary or impossible for the parents to do anything about them. It seemed that at nest Delta, C1's distress vocalizations were instrumental in stimulating the adult to help her young. In several other cases I have observed less obvious forms of intervention. In some cases parents, who were otherwise just standing in the nest, picked up and tossed grassy nest material on top of their young as they began to fight. In other instances, fighting appeared to stimulate the adults into immediately feeding and hence placating the combatants. Once I even saw an adult peck C1 on the top of the head immediately after C1 pecked C2; the effect was to bring C1's aggression to an immediate halt.

If two male or two female eaglets comprise a brood, the younger eaglet is usually subordinate to its older nest mate only for the first three or four weeks of life. By the end of that period the age effect on the difference in size between siblings is negligible.[1] The closer the combatants are in size, the less frequently overt aggression is used. Threats replace pecking and biting; for example, the mouth is held wide open and the aggressor lunges toward, but does not strike, its sibling. If a female is hatched first and a male second, the male is subordinate because of his smaller size. I have seen many fish brought to such broods; it is only after the female is satiated that the hungry male has a chance to feed. The reverse situation of male hatched first and female second is rare. I observed only 1 case in 37 broods. Even though there was an overall nestling sex ratio of one to one, the order in which the sexes hatched was not random; 63 percent of first-hatched eggs were females, and 68 percent of second-hatched eggs were males. A sex bias in hatching sequence has only recently been discovered in birds.[3] After studying the growth and competitive abilities of each sex, I concluded that it would be advantageous for Bald Eagles to avoid the male-first female-second combination. Because males develop faster than females, the size difference between siblings in the first week or two of life would be so great that the second-hatched chick would have little chance of surviving. More young can thus be raised after manipulating the sex ratio to reduce the size disparity between chicks in a brood.

In addition to influencing the probability of a chick's survival, sibling competition also has a profound effect on several aspects of development.[4] Weight gain of second-hatched chicks is slower, and the period of maximum growth is delayed compared with their sibling. The emergence of the second down and the wing, tail, and body feathers are also retarded in subordinate birds. The age at which feathers first appear can thus be quite variable, depending on the health of the chick. Siblings can appear to be farther apart in age than they really are. No doubt that is why Charles Broley, whose extensive experience with eagles was primarily with banding the young, thought that Bald Eagles laid their eggs weeks apart.[5] In fact, usually only a few days separate successive eggs.

Sibling competition determines not so much the absolute rate of development as the rate of one eaglet relative to its sibling. A suppressed second-hatched eaglet may grow faster than a first-hatched bird from another nest. The quality of the nesting habitat appears to be the overriding influence in determining how well, in absolute terms, the chicks will grow.[6] We have found that the growth of the chicks, both in overall weight gain and in feather development, was correlated with the total weight of prey delivered to the nest by the parents. As one would expect, the more food brought to the nest the faster the chicks grow. The number of fish delivered was not a factor, because the size of prey varied so much among nests. The average fish brought to one nest weighed only 288 grams (less than $2/3$ pound), whereas at another it was 760 grams ($1 2/3$ pounds).[6]

At nests containing two chicks, a fish is brought every 2.3 to 4.4 hours throughout the daylight period, averaging about one fish every 3 hours. Less food is delivered if there is only one chick. Oddly though, the rate of food provisioning is the same when chicks weigh 90 grams as it is when they weigh 5,000 grams.[7] If the amount of food delivered late in the season can sustain large chicks, then there must be an overkill of prey when the eaglets are young. This seems to be true. In the first ten days after the eggs hatch, the weight of fish in the nest is usually more than twice the weight of eaglets. We have seen as many as 24 different fish in a nest, totaling 7,800 grams (more than 17 pounds), during a single visit. Much of the excess just rots or is consumed by maggots. In the last few weeks, however, one generally finds little or no food in the nest, for it is gobbled up as soon as it is delivered.

We have referred to food as fish because birds and mammals are uncom-

A female Bald Eagle gull-wails as her mate lands on the nest with a fish. Note her hunched posture (Gary R. Bortolotti).

mon prey at Besnard Lake. From our repeated visits to nests and observations of prey deliveries, it seems that each breeding pair will bring in about one duck per year. Rarely are remains of a rabbit or squirrel found in a nest. Perhaps it is the novelty, but the chicks seem to be disproportionately excited about eating birds (preplucked by the parents). The young squabble over every morsel of a duck, including the webbed feet, even if there is plenty of fish in the nest. I suspect that it is the color of the meat that is so stimulating. While climbing into a nest one day I noticed that the eaglet attacked the red rope I was carrying. Later I did a few experiments. I offered chicks pieces of red and yellow plastic flagging tape. Red was vigorously bitten, but yellow was always ignored. Although it is probably not surprising that a carnivore should be stimulated by the color red, the results are a little odd given that Bald Eagles primarily eat the pale flesh of fish.

The adult male provides most of the food for the first week or two after the eggs hatch, but for the most part both sexes brood, feed, and provide for their young. For many raptors, the Golden Eagle being a good example, the role of the male is almost entirely one of provider.[8] He brings food to the nest while the female feeds and broods the chicks. If Bald Eagle partners do not exchange roles frequently enough, the bird on the nest may call to its mate for a relief of duties. At one nest I watched in 1976, the female had been caring for the chicks for several hours and had repeatedly called to her mate; she would look in the direction where he was perched not far away. Finally, as if tired of waiting, she took off, circled around, and dived at the male, almost knocking him off the perch; she landed on a tree nearby while he went straight to the nest without a sound.

Males, being smaller, are the subordinate sex of a breeding pair. It is common for males to have their prey taken from them by their mates, while on the nest or away from it. A female's response to a male bringing food to the nest is quite variable. Depending on her motivational state, she may either ignore him or be aggressive. A weak response is for the female to fluff out her feathers, perhaps to make herself appear even larger. A strong reaction, and certainly the most aggressive of eagle displays, is accompanied by a loud, long, drawn-out cry like some calls of large gulls, a vocalization I call a gull wail. Before the male even lands on the nest, the female leans forward, lowers her head, fluffs out her feathers, and begins to call. After the male has landed, the wails may be ear-piercing. The female frequently does not change posture or be silent until the instant that he leaves the nest. All degrees of response between mild and extreme exist. Typically, the female gull-wails only when the male has a fish. I have never heard a male gull-wail to a female, and only once have I seen the behavior directed toward another species, an Osprey who was dive-bombing a perched adult. In many cases the wails seem to drive off the male. At other times he may remain on the nest but the female will take his fish away from him. Even if the female is feeding on a big fish when he arrives, she may rush over to him, gull-wail, and take his fish. Sometimes the female is perfectly content for the male to feed the chicks and himself, but at other times, perhaps when she herself is hungry, she demands control over all food.

Given the female's dominant role, there are times when the male is not allowed to eat on the nest. The male adjusts by feeding immediately after catching prey. Observations of food deliveries to nine nests on Besnard Lake showed that if the female was on the nest when the male arrived, he had already partially consumed almost one third of all his fish. However, if the female was off hunting away from the nest, the male would previously eat only one fish in ten. The presence of the male had no effect on how often the female ate her prey before coming to the nest.

In the early period of relative food abundance, the chicks do little more than eat, sleep, and occasionally fight. To feed, the young raise their heads and take tiny pieces of fish from their parent's bill. Occasionally the male feeds one chick while the female feeds the other, or the male passes food to the female, who in turn feeds the young. One cannot help but be impressed by the delicate and tender way in which the adults feed their tiny young, especially when one has seen how mature eagles can tear off

Opposite page, top: A male Bald Eagle lands on the nest with a fish. The female hunches over and fluffs out her feathers. Middle: A few seconds later, the female is still fluffed out. Bottom: The female takes the fish from the male. Above: The male takes off, and the female's feathers return to normal. The entire sequence of events from the male's arrival to his departure lasted one minute and forty seconds (Gary R. Bortolotti).

and swallow enormous chunks of flesh for themselves. As the parent bends over to offer food to a chick, a clear fluid sometimes drips from the nasal cavities onto the upper mandible and then onto the food. The fluid may be of some nutritional importance to the young.[9] At Besnard Lake, feedings take place about once every two hours, but not all chicks in the nest are fed, or fed equal amounts, each meal. Usually the oldest nestling is fed first and fed the most. During some meals, the first-hatched eaglet pecks its sibling into a submissive posture, with head bowed low in the nest. As long as the younger bird holds its head down, it will not be fed.

There is almost always at least one adult on the nest for the first two weeks of nestling life.[10] During that time, the chicks cannot regulate their own body temperature and so must be kept warm by the adults.[11] Quite commonly, a chick will rest at its parent's breast, and in the first ten days at least, the adult may pull dried grass on top of the little bird. Young chicks may also be covered with grass when an adult is about to leave the nest.

Eaglets are never left unattended for long. Though they may not be on the nest proper, the adults are never far away. Other raptors or ravens might prey on young chicks, but we suspect that such mortality is rare. The general freedom from predation may be why Bald Eagle chicks do not respond to the alarm calls of their parents (by crouching low in the

The female stands guard over her 23- and 22-day-old chicks, who lie sprawled out and asleep in the nest (Gary R. Bortolotti).

nest, for example) as many nestling birds do. The very young eaglets also do not react to humans in any particular way except to occasionally chitter in distress. Tiny chicks will bite at a finger held in front of them, but that is unlikely to be defensive behavior.

At about one and a half to two weeks of age, most young weigh between 500 and 900 grams (1 to 2 pounds) and so need less protection from enemies. However, they are still vulnerable to the elements. Although cold temperatures are rarely a problem, heat stress can be a danger. Even at northerly latitudes the adults must spend a proportion of a sunny day on the nest to provide shade. Many times I have seen chicks dive under their parent's belly as soon as the older bird arrives on the nest. An adult often spreads its wings slightly and typically turns its back to the sun, presumably to cast as much of a shadow as possible.

Between the ages of 18 and 24 days the chicks gain 100 to 130 grams (about 4 ounces) per day, more weight than at any other stage of their development (Appendix 7). Some birds on Besnard put on more than 200 grams (7 ounces) in a single day. This period in chick development is my favorite if for no other reason than its delightful, comic effects. The chicks even look like clowns. Gawky and pot-bellied, with tufts of down sticking straight up on their heads, they have lost the cuteness of the newly hatched. They are largely covered with their second coat of down; it is dark gray and much woollier than the first. The feathers that will constitute the plumage of juveniles have not quite begun to emerge. Until now, the young have only squatted or lain down in the nest and waddled about by shuffling their huge legs in front of them. They soon begin to stand up, albeit weakly. It is interesting that the first standing motions, as well as much of the movement around the nest, are associated with defecation. The young always back up to the rim of the nest and, to use falconers' jargon, "slice" — feces shoot out with remarkable force some distance into the forest. After the breeding season, one can often tell if a nest has been used by the ring of "whitewash" around the base of the nest tree. Standing just before slicing is common long before the birds can stand to walk. Adults, by the way, almost never slice while on the nest.

In the middle third of the nestling period (weeks four to seven), the appetites of the eaglets are enormous. Gone are the days when the eaglets patiently wait for morsels of food to be passed to them. The young birds now grab at their parent's bill, pushing and shoving against each other and the adult. The eaglets won't even wait until their parent has torn the fish up for them, and enormous chunks of fish may be snatched away and gulped down. It is surprising to see how much food can be eaten at a time. Fully grown Bald Eagles can consume about 900 grams (2 pounds) of flesh in a single meal.[12] Large nestlings can eat nearly as much. Not only does the stomach fill with food during a feeding but so does the crop, a storage pouch on the throat that can hold the equivalent of about 15 percent of the bird's weight. A full crop shows as a conspicuous bulge in the throat. Gradually, as food is digested in the stomach, the crop empties.

Self-feeding begins at about the sixth to seventh week of age, but it is a long time before the eaglets become proficient. They will nibble up and down a whole fish carcass but for most of their nestling life do not have either the strength or the know-how to tear into it. Sooner or later their hooked bills find the fish's mouth, gills, or sometimes even the anus. Finally, after many unrewarding tugs, they pull off some flesh, and self-feeding begins. The head area of the fish is the easiest and most convenient place to start to tear up a carcass; perhaps that is why so many fish-eating raptors eat the head first and thus bring headless, rather than tailless, prey to the nest.

By eight weeks old the eaglets are quite capable of standing up and walking about the nest. They also strike at the nest material with their feet; a quick stab at a stick may be a precursor to prey catching. However, the young are far from being predators. On one memorable occasion, an adult dropped off a sixteen-inch sucker at the nest and then hopped onto a large branch of the nest tree. The fact that the fish was still alive did not go unnoticed; the two nestlings cautiously approached the beast, for they saw

These ten-week-old eaglets whine and fluff out their feathers when they see their parent coming home with a fish (Gary R. Bortolotti).

its mouth and gills moving. The bravest youngster leaned slowly forward and barely touched it. The fish flipped and rolled over; both birds drew back startled. After a short pause, both young moved closer. Cautiously, they touched the fish; again the fish flipped over. The eaglets stepped back in surprise. On their third inspection the sucker gave a last desperate twist and sent itself over the edge of the nest and into the bushes forty feet below. A pound and a half of fresh meat, quite a meal, was lost forever. The adult, who had been watching everything, quickly turned about-face on its perch as the fish went over the side. Both eaglets waddled to the rim of the nest, leaned over the edge, and just stared downward for the longest time. If ever an anthropomorphism was appropriate, this was the time; the eaglets seemed truly amazed by the experience.

The point at which nestlings are capable of feeding themselves marks the beginning of another major phase of development. Eaglets about sixty days old are well feathered and have achieved more than 90 percent of their mature weight. The female nestlings may even be larger than their fathers. The young are increasingly aware of their surroundings outside the nest and may react to the presence of strange eagles or other birds nearby. They are also aware of the adversarial nature of humans at the nest. After they are 55 days old, and sometimes before, there is a good chance that if a human climbs the nest tree, the eaglets will jump off the nest in an attempt to escape. Otherwise, the chicks respond to a human in their nest by backing away, hissing, and adopting the spread-eagle posture (wings held out far to the side). Presumably this is a defense against predators. Twice we have seen fresh claw marks left by black bears that have climbed right up to the edge of a nest. As the chicks were not taken, the bears may not have been able to get around the rim of the nest or were scared off by the nestlings or adults.

When people hear that we climb into eagles' nests, the most frequent question they ask is "Do the parents attack you?" The behavior of the adults is actually quite variable and ranges from timid to bold. Some adults swoop at us and hit branches of the nest tree, but most pairs just circle nearby and call excitedly.[13] Of the thousands of Bald Eagle nests that have been climbed to across the continent, there are only a few records of researchers' being hit in an attack.[14] If there is a bird an eagle bander should be afraid of, it is usually the nestling.

Some eaglets can be as gentle as lambs when being handled, but others are almost impossible to control. Usually they hiss, or, in a most swinelike manner, grunt in deep, throaty tones. Unlike most raptors, which fight almost exclusively with their feet, Bald Eagles also wield their massive bills and can deliver a nasty bite. Although an eaglet's grasp is not as powerful as an adult's, its talons draw blood nonetheless. Charles Broley found that out the hard way. One scrappy Florida Bald Eagle managed to set the talons of one foot deeply into Broley's left hand, while those on the other foot splayed over Broley's face. Charlie was uncertain exactly how he got clear of the eaglet, but it was quite a tussle. Even though bleeding profusely, he banded the bird before climbing down. After mopping up the blood, Myrtle Broley measured the distance from the talon mark on the top of her husband's head to the one on his chin—7½ inches.[15]

The young are aggressive toward their parents during the self-feeding stage, for they seem to be constantly demanding food. The nest is no longer a place where an adult, especially the small males, can comfortably loaf and feed. When a parent lands, the eaglets typically rush toward it screaming at the top of their lungs. So rambunctious are the eaglets that they attack their parents with wings outstretched, mouths open wide, and bodies lunging forward. Often, the male hardly even touches down on the nest before leaping away. Should the adult not have a fish, an overeager youngster grabs whatever it can get, usually a toe, and an adult-offspring tug-of-war ensues. More than once I have seen an adult eagle pushed right off the nest and sent scrambling for a foothold down the side. When an adult does succeed in staying on the nest, the loud, high-pitched, and drawn-out whines of the youngsters may continue as long as the adult remains. When a nestling secures food from an adult, the eaglet may turn completely around and crouch over the food with wings held out to the side ("mantling" is the falconer's term for this), effectively hiding the food. Eaglets mantle in response to the presence of their parents, a sibling, or rarely to a human in the nest.

The adult female will at times remain on the nest after delivering a fish. Probably because of her size, she is less physically harassed by her offspring than is her mate. She may even try to feed the young; however, the offerings of food and the responses by her young are not at all like those that were typical just a few weeks previously. Everything happens in slow motion. The female will tear off a piece of fish and bend over toward a nestling. The eaglet, however, remains distant, often holding its wings out at full extent. If the eaglet approaches and takes the morsel, it will do so with the utmost care. Typically, little of the fish is consumed, and the adult soon departs. The male rarely even tries to tear at his fish; he usually doesn't get a chance to, for it is snatched away as soon as he lands. It is not surprising that the frequency with which males bring partially eaten prey to the nest triples after the chicks become aggressive. Whereas at one time they could feed freely on their prey at the nest, the adult males must now eat first if they are to eat at all.

While watching nests from the blinds, I was usually warned that an adult was on its way home with a fish long before I actually saw it; the nestlings would spot their parent first and start their loud, screaming whines. Adults sometimes return from a fishing sortie and head directly to the nest but at the last moment veer away, circle the area, and even perch for a while before dropping off the fish. This has been interpreted by some observers as the way adults entice their young into making their first flight. More

The eaglets, with wings held out to the side, sun themselves (Gary R. Bortolotti).

likely, the parents are just reluctant to go to the nest, given the response that awaits them.

In the last two or three weeks before fledging, the flight feathers have grown enough that the eaglets, now about 95 percent of their mature size, can propel themselves across the nest in short flights. The young birds flap, jump, and generally move about a great deal in the final two weeks before fledging, and they spend increasing amounts of time on support branches of the nest. Males are noticeably more active in this respect than females, as is typically the case for raptors. About one in seven eaglets fledges prematurely, falling or jumping from the nest tree before it can fly, and is forced to spend a few days on the ground. Some birds land in the lake and have to swim; awkward as it may seem, they row themselves ashore with their wings. Males make their first flight at about 78 days of age (range 68 to 84), and females usually fledge later, at about 82 days (range 75 to 88).[16]

The first flight of an eaglet can be impressive. The young birds flap and glide reasonably well, but landing is a different story. C2 at nest Y was a good example. One afternoon as we boated through the territory we noticed C1 alone atop the nest. We soon spotted C2 perched midway up an aspen about fifty feet downshore. C2 was our youngest fledgling, only 68 days old. Early the next morning we set up a blind; C2 was on the same

An eaglet jumps and flaps in the nest in preparation for its first flight (Gary R. Bortolotti).

limb. For the next seven and a half hours we listened to C2 whining constantly but otherwise doing little else. His parents ignored him and flew directly to and from the nest. At 12:20 the adult female landed on the nest with a fish, which C1 ate immediately. C2's whines grew louder. Apparently he had had enough of independence and so took off over the lake. Flapping slowly but strongly, C2 cruised past the nest, banked toward shore, and crashed into the canopy of a tree. We could hear the slapping of wings against branches all the way to the blind. When it was over, C2 was hanging upside down from a limb and holding on with just one foot. From out of nowhere the adult male flew in, calling excitedly, and soon perched on a spruce directly above his clumsy offspring. The adult female took off from the nest, circled above C2, and joined in the chorus of cackles. C2, rather calmly, just hung there, occasionally looking from side to side. After three minutes, C2 released his grasp and crashed to the ground. Both adults were in the air circling and calling. We couldn't resist intervening. Fortunately, C2 was alive and well, and so we put him on a big rock onshore in front of the nest. The next day we found C2 perched atop a small spruce near the nest, his crop bulging from a recent meal. He held his bill high in the air as we passed; we definitely felt snubbed.

10. Whither the Wind Blows

Now rising high in air to sweep
In circling rings the upper deep.
Isaac McLellan

Jon—

The Missouri River, just north of Bismarck, North Dakota, November 28, 1975.
Peculiar mounds in the prairie surround me, the remnants of an ancient
ditched Mandan Indian village. I am standing on a hill overlooking the
Missouri River. Except for myself, my wife, Naomi, and our two-year-
old daughter, Pauline, the land is deserted. It is easy to imagine how it
had been when the village was there, positioned with a marvelous view
of the wide, winding river below, both upstream (north) and downstream. Then, the river was a major artery for travel. Indians, explorers,
and fur traders regularly paddled past. In 1805 a flotilla of canoes
belonging to the party of Lewis and Clark may have stopped. On the
other side, there is a narrow floodplain, and then hills rise gradually into
the distance. Beyond the hills, the rolling country goes back toward the
land where Teddy Roosevelt came to herd cattle nearly a century after
Lewis and Clark. Behind us, on this side of the river, flat open prairie
stretches to the eastern horizon. We have come based on a hunch that
eagles follow the Missouri River on migration.

Yesterday, three adult Bald Eagles were perched one to three miles
apart along the river north of this spot; one immature circled locally, but
there was no migration. Today we arrive at 10:45. The day does not
appear particularly auspicious. The sky is overcast. There is a light snow
falling. The wind, however, is from a good direction, the northeast, at
about eight miles per hour. As I wait, I wonder what the Mandan
Indians who lived here day after day and season after season could have
told us about migrating eagles. A lot, I suspect.

Five minutes after eleven, I see the first bird. An adult Bald Eagle
flaps, then glides, flaps, then glides, going southeast just above the trees
on the west bank. The breeze from the northeast provides weak updrafts

for the eagle; the updrafts are not always strong enough and periodically the eagle flaps. A minute later, a second eagle follows the same path; a minute after that, a third. Oh, what a day!

But moments later the wind decreases. An adult flies into view, then perches along the shore to look for fish. For two hours no more eagles pass. Near one o'clock, another adult flaps steadily south. The air is calm, and the bird, receiving no assistance from the wind, has to flap continuously. Twenty minutes later, another and then another eagle appear. Each one flaps steadily down the center of the river, about a hundred feet above the water. Then, just before one-thirty, there are two more eagles. These two now find thermals. I watch in admiration as the first and then the second circle gradually higher and higher. At the top of the thermal, they start a long glide south along the river and continues until they are out of sight.

I wait another hour, but the migration, at least for today, is over. I glance again at the Indian village. Each mound was once the site of a hut. Around the whole village is a large defensive ditch. With the European fur traders and explorers came smallpox. The Mandans, being on this major artery, were one of the first tribes to suffer the ravages of the European scourge. Many died. The village shrank. As I look closely, I notice a smaller second ditch inside the first — all that was needed to enclose the village after the smallpox epidemic. It was, sad to say, not the last time the Indians of this part of the Missouri River were affected. With time the Mandans passed into history and became extinct. I think about the eagles. It is 1975. The breeding range of eagles has shrunk drastically in the eastern half of North America. Would the slowly shrinking circles on the map representing eagle breeding areas be like this shrinking village, a harbinger of worse to come?

The Saskatchewan River, south of Saskatoon, Saskatchewan, April 1981. Along a different river and under different conditions (this time in the spring) Naomi and I again stop to look for eagles on migration. During our first few days we found eagles roosting along the river south of Saskatoon. Like the birds on the Missouri in the fall, they usually perch at intervals of a mile or two along the river. We have tried to find the route these birds take when they head north in the morning. So far we have been unsuccessful. They usually do not follow the river as they travel; observations to the east of the river have drawn a blank. This morning, on the west side, we travel within the river floodplain on the low road to Pike Lake. It is a sunny day, a good sign, for eagles in spring do not seem to move on overcast days (different from the fall, when flying on overcast days is common). However, there is a light wind from the north. An eagle wanting to migrate today will have to go into the wind. I guess there will be little movement.

Opposite Beaver Creek, which enters on the east side of the river, we turn west, climb the escarpment out of the floodplain, join the high road to Pike Lake, and continue south. As we pass along the edge of the valley, we spot an adult Bald Eagle circling low over the field beside the road. I brake the van. We watch. The bird flaps west across the road and finds another thermal. This is a stronger one. The bird climbs. Circling, it rises to 300 feet. As it does so, it drifts south, because of the wind, then from the top of the thermal it glides north. The eagle quickly regains the distance lost when it drifted south and continues on. A quarter mile north it finds a third thermal; however, this one is too weak.

Once more the eagle crosses the road. A thermal on this side, on the crest of a small rise, is stronger. The eagle climbs to 600 feet. Then in another cone of hot air, it rises another 900 feet before heading northeast. It goes half a mile in a glide before needing another thermal. We follow it. The same pattern is repeated again and again. Using the local grid of roads, we find we can keep the eagle in view across mile after mile of prairie and partly wooded land. At one point, the eagle cannot find a thermal at the bottom of a glide and so flaps for half a mile before finding another. Slowly and steadily it moves northeast, away from the river. The land is just a little more treed and has a little more relief than anything to the east or west. The eagle travels quite quickly in a glide, one we clock at 45 miles per hour, but its frequent use of thermals and its southerly drift in each thermal, caused by the north wind, means that its average speed is relatively slow, not quite 15 miles an hour. In seventy minutes it covers only 17 miles. In the end we lose the bird into the high blue sky. For the next two days we track many other eagles along this route — an aerial highway, one not marked on any map but used by generation after generation of eagles.

Knowledge of Bald Eagle migration really begins with the work of Charles Broley. In 1939, when he retired from his position as a bank manager in Winnipeg, Charles Broley and his wife planned to spend much of their time at their summer cottage near Delta, Ontario. They decided, however, to spend their first winter in the Tampa Bay region of Florida. On their way south in the fall, they stopped at a meeting of the American Ornithologists Union in Washington and encountered Richard Pough, a noted naturalist with the National Audubon Society. Pough persuaded Broley to take four eagle bands with him to see if he could find a nest with young; he was concerned that the Bald Eagles of Florida were being shot and was also interested in finding out something about their movements.

Broley found his first nest soon after arriving in Florida. By mid-January he had put the first four bands on nestlings and wrote for some more. A bird Broley banded near Tampa Bay on January 28 was shot near

Columbiaville, New York, on May 8. The recovery was exciting; it was the first evidence that Florida birds migrated north after they fledged.

The results spurred Broley onward. Soon he was banding more than a hundred eaglets a year. Recoveries in Georgia, North and South Carolina, Virginia, Pennsylvania, New York, Massachusetts, Connecticut, Maine, Quebec, New Brunswick, Nova Scotia, and Prince Edward Island showed that most birds went north along the east coast after they left their nests. A few eagles went farther west (Kentucky, Ohio, Indiana, Illinois, Arkansas, Minnesota, and Manitoba). The findings were a major contribution to ornithology in North America.[1]

Broley also banded in Ontario. These birds followed a more usual migration pattern, going south or southwest from their breeding grounds in the fall. After Broley died, in 1959, major efforts to band eagles got under way in the mid-sixties, with Sergej Postupalsky in Michigan, Chuck Sindelar in Wisconsin, Jim Grier in Ontario, and Doug Whitfield and myself (J.G.) in Saskatchewan. After several years of work, we began to realize that the proportion of banded eagles that were later recovered was much lower than expected based on Broley's data. Although part of the reason was that fewer eagles were being shot (a good development), we became concerned that eagles might be removing the bands. In the spring of 1974, at Quandal, in northern Iowa, Postupalsky, Grier, Sindelar, and J.G. got together. They assembled enough evidence to suggest that Bald Eagles could remove the bands then being used, a lock-on variety in which a flap of aluminum was bent over to secure the band in place. On a recommendation from the group, the banding office changed to a band with two flanges of aluminum riveted together so that it could not possibly be removed by the birds.

A perusal of the data collected by those banders showed that the general pattern of eagle migration from the various states and provinces was southward in winter. Eagles from Michigan nested a little earlier than those farther north; as the young fledged earlier and had more time before being forced to go south, they sometimes wandered. One bird even went to James Bay in northern Ontario. Eagles farther north, in Saskatchewan, with less time to spare, appeared to head directly south. Movements of eagles released in New York were even more random than those of nestlings from Michigan, with only a slight southward tendency. Eagles from the Chesapeake Bay show a general scatter, except of course that they do not move east into the Atlantic Ocean. The results of all these studies suggest that although Bald Eagles are migratory in parts of their range where their habitat is inhospitable in winter, they move less, or more or less randomly, in other areas where there is an adequate food supply year-round.[2]

At about the same time that better bands were being produced, other advances in studying eagle movements were being made. Michael Kochert, working in Idaho, started putting colored vinyl tags on young Golden Eagles. The tags were visible for up to half a mile with binoculars, thus

Dispersal of young Bald Eagles from Besnard Lake, Saskatchewan, and near Delta, Ontario (top figure), one release site in New York state (left figure), and Chesapeake Bay (right figure). A line connects the point of origin to the point where a bird was sighted (if color-marked) or found dead. (Data from Gerrard et al. 1978 and unpublished data; Broley 1952; Nye in press; and Cline 1986.)

A ten-week-old eaglet with two vinyl color markers wrapped over its right wing. Only the marker farthest out on the wing is clearly visible (Jon M. Gerrard).

making it possible to identify an individual from a long way off. Dan Franzel in Minnesota and Doug Whitfield and J. G. in Saskatchewan applied them to Bald Eagles. One of the first things that was learned concerned the movements of the young eagles immediately after they left the nest. Most fledglings stayed within a mile of the nest for six to eight weeks.[3] It was interesting that a number of the eaglets were found on the ground during the initial weeks after fledging. The young birds appear to find it easier to perch on the ground until they are more skilled at flying and landing. Young from the nest we monitored using time-lapse photography returned periodically to the nest either to visit or to feed on food left by the adults. Throughout this period, the adults continue to bring food to the young; the eaglets have a plaintive food-begging call to enable the adults to find them even if they are perched low and camouflaged well. The movement of the young away from the nest revealed a tendency to

Juvenile Bald Eagles return to their nests from time to time in the first few weeks after fledging (Gary R. Bortolotti).

drift downwind. The direction taken was also influenced by the shoreline, since birds tended to follow the forest edge rather than move overland. Eagles 21 weeks after hatching (7 to 9 weeks after fledging) were no longer found on the lake as frequently, and one was seen 350 miles to the southeast.

During October, November, and December there were many more reports of sightings of our color-marked birds on migration. Although each report had to be checked carefully, and many were rejected because of uncertainties, the colored wing tags produced in a short time much more information on eagle movements than had been possible with banding. Saskatchewan eagles were found to migrate south in a relatively triangular distribution, with important routes along the Missouri River in the east and through eastern Montana to the west.[3] Additional valuable information was also obtained. For example, an eagle from Besnard Lake was sighted several winters in a row in the same location, establishing that a bird will return to the same wintering area year after year. Resightings of eagles on the breeding grounds showed that many returned to the lake where they had been raised, first to spend the summer and later to breed. The occasional bird did wander some distance; one eagle from Besnard Lake was seen in summer about 250 miles to the northwest.

Color-marking has been successful, but since there is a limited number of colors and a good continent-wide color-marking scheme was not put in place, the overlapping use of colors by different researchers soon made it a less useful technique. More recently, radiotelemetry has been employed to investigate movements. The initial transmitters developed for eagles had to be mounted on a cumbersome backpack harness. Some information was obtained with these awkward devices, in particular by Jim Harper in Minnesota.[4] From September 17 to September 21, he followed one young eagle from northern Minnesota to near Glad Valley, South Dakota, a distance of 410 miles from the nest; it covered an average of 95 miles per day. He then tracked another eagle from northern Minnesota. This bird left the vicinity of its nest September 22 and wandered rather randomly until October 11 (it was then about 50 miles southwest of its nest), when it appeared to begin migration. For two days it headed steadily south but, instead of continuing, made a long circle covering 560 miles to arrive back where it had been the morning of October 11. The results suggest that young-of-the-year usually migrate independently of their parents, and show that some birds may wander long distances from the nest site but then return before finally migrating south for the winter.

Following the development of tail-mounted transmitters, which were lighter and less likely to interfere with eagle activities, Al Harmata in 1981 radioed four adult eagles. He captured them on their wintering grounds in the San Luis Valley, Colorado, and followed them north.[5] One in particular that he wanted to follow was paired. The male headed north without its mate, got caught in a storm, and then came back to its mate for a day before starting out again. This time it went to the middle of Wyoming before Al lost track of it. (With a following wind, the eagle left Al behind even when he pushed his truck up to 90 miles an hour.) In searching for the eagle, Al found one of the other radioed birds and followed it north all the way to Chachukew Lake, a small lake in northern Saskatchewan. There the bird nested and raised an eaglet. In following that eagle, and others the next year, Harmata was able to show that most eagles start their migration in March (one started as early as late January, and one as late as early April), that eagles average about 30 miles per hour and travel from 90 to 270 miles in a day, and that the routes often follow prominent physiographic features such as deep canyons, north-south mountain ranges, and rivers.

Al Harmata's success led others to attempt similar tracking. Riley McClelland set out to follow eagles trapped at McDonald Creek in Glacier National Park, Montana, where eagles feed during October, November, and December on large numbers of spawning kokanee salmon.[6] Riley McClelland, his wife, Pat, and Leonard Young followed a number of eagles, first south and southwest to where they wintered, including regions of Idaho and Oregon, then north in the spring through Alberta to breeding areas in the Northwest Territories.

Some interesting observations of eagle movements have been made in other regions.[7] In Nova Scotia Bald Eagles appear to stay largely within the province, but a few birds have gone as far as Maine. Some Bald Eagles raised at nests in Maine (20 to 50 percent) will similarly stay there in the winter, though a considerable proportion (30 to 60 percent) wander southwest along the Atlantic Seaboard to Massachusetts, Connecticut, Maryland, or even South Carolina. In Wisconsin, Chuck Sindelar reported 28 band recoveries: 16 were in Wisconsin, 2 in Iowa, 2 in Minnesota, 2 in Oklahoma, 1 each in Mississippi, Missouri, Illinois, Kansas, Arkansas, and Colorado. In the Greater Yellowstone Ecosystem some of the young eagles move to winter in the west coast states of California, Oregon, and Washington, while others remain in the Yellowstone area. Most of the adults remain in the region for the winter, though there appears to be some movement out of Yellowstone Park into nearby areas.

Along the west coast, radiotelemetry of eagles feeding on salmon in the Skagit River Valley, Washington, has shown that eagles travel considerable distances up and down the coast, with some of the wintering eagles originating from central British Columbia. In general it would appear that seasonal movements up and down the coast are coordinated with the spawning runs of coho, dog, and pink salmon (see Chapter 11). Eagles in interior Alaska probably migrate to the coast. Eagles on the Aleutians are believed to be nonmigratory, though there appears to have been some movement of immatures away from Amchitka Island when the food supply decreased in 1974.

The Banders

Since so much has been learned about migration as a result of banding, let me pay tribute to the efforts of two of the banders, Chuck Sindelar and Dave Evans, by describing their work.[8] Chuck, for years a laundromat operator, and Dave, a student, began as amateurs and have been banding the Bald Eagles in Wisconsin for many years. Usually they are in the field for several weeks and have all the appearance of real backwoodsmen, for frequently they have to walk through thick woods and marshes to get to the nests. Helen Cummings, an elderly lady who lives in northern Wisconsin, describes her first visit from Chuck and Dave.

> On a rainy June 7th I was standing at the door of my cottage . . . when a Volkswagen pulled into the turnaround. [Chuck and Dave] had been out in the field for more than a week. It's a rugged life. Working as volunteers on their own time and money, they utilized every daylight hour and in June the sun rises early and sets late. . . . Every time Dave moved he moaned, and I asked Chuck what the trouble was. He said, "I imagine he's in pain." A month earlier, they had been in Minnesota collecting eggs to be transferred to Maine. . . . Dave, unused to the audience that had gathered to witness the momentous event, fell 40 feet. He . . . bruised ribs badly [but] climbed again

two days later. After going over the maps and planning the schedule for the next day, they got to sleep. . . . Just before they were to leave early the next morning, they asked if I wanted to go along. I had my boots on just in case, and I went out without even closing the door.

Helen soon became a gaboon — a helper — and has joined Chuck and Dave every year since. She continues:

> If I have to do housework, or shovel a little snow, or walk a couple of blocks, I get shortness of breath, pains in the chest, and arthritis in the knees, but when it comes to visiting eagle nests, I can carry canoes, climbing irons and ropes. I can jump from hummock to hummock through the swamps, and wade up to my knees in mud all day, accompanied by mosquitos, deerflies and woodticks. I feel like a million bucks when it is over. . . .
>
> Unless you've been there, I don't think you can imagine how breathtaking it is to watch a man climb a tree, sometimes well over 100 feet high, and then make his way around that great bulky structure of a nest. Dave wears the spikes worn by telephone linemen. He relies on his hands and arms for safety. He uses a mountaineer rope 150 feet long with which he can tie himself in if necessary. If the nest is sturdy and supported by strong, live branches, he gets into it. Otherwise, he uses a more hazardous method. He ties in under the nest, and with a 'grabber', which is a long heavy wire with a hook on the end, he pulls the eaglets out to where he can band them.[8]

General Aspects of Eagle Migration

Flight Patterns and Migratory Routes
In both spring and fall Bald Eagles are remarkably consistent in starting their daily flight between 10 and 11 A.M. if the weather is reasonable. That is usually the time when good thermals become available. In general, thermals are better in spring than in fall, and turbulence in the atmosphere is greater. As a result thermals are important to spring migration. Occasionally, a following wind and a row of thermals will allow an eagle to soar continually downwind in big circles without using any significant amount of gliding. In the fall eagles may go south on snowy overcast days without thermals. On such occasions, eagles may choose to glide in updrafts off the slope at the edge of a river valley. Flapping flight, because of its high metabolic cost, is avoided whenever possible. Often, in fall eagles must use a combination of the available thermals, slope lift along a river or line of hills, and long glides, sometimes with flapping between each lift. A notable example of eagle use of slope lift is along the Kittatinny Ridge — originally "Kau-ta-tin-chunk" ("endless mountain") in the tongue of the Lenape Indians — which stretches, almost without a break, like a winding snake from Maine to Georgia. Information on migrating eagles has been gathered at various locations along the ridge, particularly at Hawk Mountain in Pennsylvania.[9] Here, as late as 1952, more than a hundred Bald Eagles were seen each fall, with a peak in September; they were and are still largely

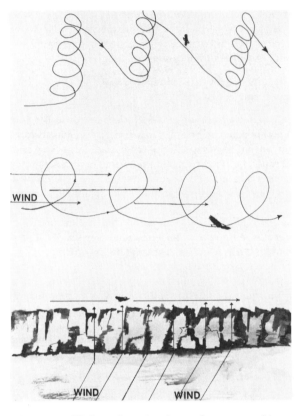

Three types of flight used on migration: eagles may ascend in a thermal and then glide down (top), circle steadily down a street of thermals (middle), or use the rising air generated when wind sweeps against a cliff or other raised feature of the landscape (bottom) (Naomi Gerrard).

Florida-bred migrants returning south. In 1953 the total was sixty. In the years since, through 1980, the average has been only forty.

Updrafts along the ridge of the Mississippi and Missouri river valleys in the west, similar to the Kittatinny Ridge, are used by many Bald Eagles each spring and fall.[10] At Eagle Valley in Wisconsin on the Mississippi River, 1,455 Bald Eagles were counted migrating between September 12 and December 19, 1981.

Eagles choose their migratory routes to take advantage of thermals, updrafts, and food sources. In Saskatchewan, much of the migration appears to follow upland areas. In autumn, the birds frequent stopover points near lakes or rivers. In the spring much more of the migration is cross-country without any obvious orientation to natural landmarks, though sections of a migratory route may follow rivers or mountain ranges. Indeed, when birds cross the plains almost any irregularity in the terrain may serve as the local migratory route.

Researchers have speculated that eagles might occasionally migrate at night (using slope lift along river valleys or thermals off lakes in the late fall, when water bodies are warmer than the surrounding land, which cools rapidly overnight), but good evidence has never been obtained. Instead, radio tracking of eagles suggests that all migratory movements occur during the day between 8:00 A.M. and about 6:00 P.M., with most eagles not starting till 10:00 A.M.[5]

A comparison with the Rough-legged Hawk is instructive. The wing loading of eagles is much greater than that of the Rough-legged Hawk, and eagles need stronger thermals and updrafts. However, once eagles are up in the air, they can travel much farther than the hawks without as much additional lift. Rough-legged Hawks are less selective in choosing their migratory routes and will cross flat, open country. Eagles in contrast almost always pick areas with more relief, such as river valleys and hills.[11]

Feeding and Stopover Sites
On their way south in the fall, Bald Eagles will stop off at sites where there is food. Glacier National Park, mentioned earlier, is a good example. Smaller numbers stop over at innumerable lakes and wildlife refuges in southern Manitoba, Saskatchewan, and Alberta and the northern plains states. Feeding may be done first thing in the morning. Alternately, and more usually, as we have observed along the Mississippi River, an eagle will start migrating and continue until it finds food. After feeding, the eagles usually then find a loafing perch. At dusk, they will seek a nearby roost. The next morning the pattern will continue: starting migration about 10:30, feeding by 1:00 or 2:00 P.M., and heading to the roost by about 5:00 P.M.[12]

In spring, eagles do not necessarily feed on migration. With food generally less available in the northern plains and with the rivers still frozen during migration, eagles may travel for several days on end without feeding, even going all the way from Colorado to Saskatchewan.[5] Stopovers in spring are usually because the weather is unsuitable for migration. There are, however, some notable spring stopover points, such as the Platte River, Nebraska, and the Bear River marshes near Great Salt Lake, Utah. Several hundred Bald Eagles may visit the latter and feed on their way north. Ground squirrels in southern Alberta and carrion along the South Saskatchewan River are further examples of what eagles eat during spring migration.

Timing of Migration
The fall migration of eagles in interior North America is timed to coincide with freeze-up perhaps in part to take advantage of crippled waterfowl caught in the ice. In general, the immatures go south first; at Besnard Lake

they usually leave during October. Adults stay longer, with some staying on their breeding grounds until the lakes completely freeze over, usually mid-November for Besnard Lake. Cold fronts may trigger migration. An eagle followed by Arnold Nijssen began its southward migration coincident with the arrival of a cold front.[13] Whether this is a general phenomenon remains to be seen. At Saskatoon, about 250 miles south of Besnard Lake, the median date for the fall migration of immatures is October 25, and that for adults is November 9. The peak of fall migration through Saskatoon, Saskatchewan, occurs when maximal daily temperatures are 23° to 39°F.

Southward movement in fall, when the average distance for one adult over five days was 14 miles per day, may be much slower than in the spring, when the average was 112 miles per day for ten migratory days (see also Appendix 8)[5, 13] The slower movement southward in fall probably reflects the less suitable thermals, greater food supply, and lower urgency to reach the wintering grounds. In spring, adults that reach the breeding grounds first may have an advantage in claiming the better territories or by starting reproduction as soon as possible.

The median date of spring migration of Bald Eagles through Saskatoon is April 5 for adults and April 12 for immatures. The peak of spring migration occurs when maximal daily temperatures reach 32° to 57°F.[9] In spring, when lakes thaw, there are often fish available as a result of winterkill, caused by a lack of oxygen in the water under the ice. For adults that begin their migration in the southern states, such winterkills are frequently available in lakes in Colorado, Utah, Oklahoma, and Kansas. When the adults, which reach their nests by early April, come through the northern prairies, most lakes are still frozen. Immatures, which go north later than adults, may make more use of such winter-killed fish as they move north in a more leisurely fashion.

It is unknown to what extent pairs of eagles may migrate together. Observations of a mated pair at Besnard Lake, Saskatchewan, showed that the male left first on its fall migration.[13] Observations of a mated pair together on the wintering grounds suggested that the male left first in the spring.[5] On their breeding grounds in spring, eagles of a mated pair often arrive separately; Harmata observed one in which the male arrived four days ahead of the female.[5]

Differences Between Adults and Immatures
Two factors play important roles in the difference between immatures and adults—experience and size. Immatures as discussed in the chapter on flight are specialized to soar well, but as a result they travel more slowly than adults and are more influenced by the wind. Adults are specialized for fast directed movement; Harmata clocked one adult gliding at 89 miles per hour.[5] Adults therefore tend to have fast targeted migration, whereas immatures tend to take more time and to wander more. Inexperience of

the immatures in finding the best updrafts and thermals may also slow them down in their first fall, and their lack of knowledge of the terrain likely leads to more aimless movements. There is no evidence to suggest that young-of-the-year fly south with their parents, and a number of observations suggest they do not.

Wind

The wind has a considerable effect on migration. The first person to point this out was Charles Broley. From 1939 to 1944 almost all his band recoveries were from along the east coast. For those years the prevailing winds during April and May were from the south, moving up the Atlantic coast. In 1945 the prevailing winds in April shifted to come across Florida from the Atlantic and then swing north into the Mississippi Valley. That year six of Broley's eagles banded in Florida were recovered inland, in Mississippi, Arkansas, Illinois, Michigan, Ohio, and Pine City, New York.[1] The recoveries suggested that young eagles tended to drift with the wind. Support for an important influence of wind on eagle movements came 28 years later, at Besnard Lake, when Peter Gerrard followed the initial movements of eagles after fledging and found that the direction of their movements tended to be downwind.[3] Subsequently, additional studies of migrating eagles have demonstrated that even adults show significant wind drift, temporarily deviating from a straight line toward the target of their migration.[5, 11] The tendency of eagles to drift with the wind may explain such unusual movements as the journeys of eagles from Besnard Lake to Alaska and Maine. For immatures in their first fall, migration and movements appear to be, to a considerable extent, a reflection of whither the wind blows. For adults, wind probably has no influence on the eventual target of their migration, generally summer and winter habitats with which they are familiar. Nevertheless, the wind still strongly affects the route they take and other aspects of the pattern and timing of their movements.

11. Anything Edible: Bald Eagles in Winter

High o'er the watery uproar, silent seen,
Sailing sedate in majesty serene,
Now midst the pillared spray sublimely lost,
And now, emerging, down the rapids tossed,
Glides the Bald Eagle, gazing, calm and slow.
Alexander Wilson

Jon —

South Dakota: Fort Randall Dam on the Missouri River, November 1973. We arrived in the middle of the night, crossed the river on top of the dam, descended into the wide valley, and then snuggled in to wait for dawn. As the gray of the night cleared I looked out into the mists above the waters of the Missouri River; black shapes appeared and disappeared as they flew back and forth. What were they? The light improved. They were eagles! From time to time, one would turn over on its side, dive down to the water's surface, immediately rise steeply, and continue cruising around twenty to thirty feet above the water. Occasionally, an eagle did fly to a tree perch. The eagles were not catching and carrying fish the way I was used to seeing. At Besnard Lake, an eagle carrying a fish flies low over the water, but these birds were flying higher up. What was happening? We followed the eagles with our binoculars. To my surprise, the next bird I caught stooping to the water returned sharply to its starting height (about 25 feet), bent its neck down to its feet, and transferred something to its bill and swallowed it. We learned later it was a tiny two- to three-inch-long gizzard shad.

A few minutes later another adult eagle caught a fish, one about six inches long. As it flew to a nearby cottonwood, the adult was chased by an immature. In and out among the branches they went — I was amazed at their agility — until the adult, a bit ahead, cut sharply to one side to land. It fed for a few seconds. The immature circled around, then charged in, displaced the adult, and polished off the last remnant of the fish.

Missouri: Squaw Creek National Wildlife Refuge, November 1976. We arrived
at dusk just as a blizzard descended. In our motel we listened to the roar
of the wind. Early the next morning, we drove to the refuge. To our
surprise, the main pool was frozen solid except for a few small holes; it
was early in the winter for freeze-up. Several hundred Snow Geese
packed closely together in an area of open water barely larger than the
flock. The water had probably remained open during the cold night only
through the warmth and movement of the birds. Small groups took off,
flew around, and went to feed or returned to join the main congregation.
Elsewhere on the pool small groups of five or six ducks huddled
together. We wondered why they were separate from the larger mass of
waterfowl. Several eagles perched in trees nearby or stood on the ice. An
adult eagle left its perch, coursed low over the ice, came quickly upon a
group of the stragglers, and stretching out its right foot picked up a teal.
The other ducks flapped weakly to one side, tumbling and struggling as
they did so. None flew. Here was the answer to why the ducks were
alone. They were cripples, unable to fly. The eagle landed on the ice fifty
feet away and began tearing at its prize. Over the next two hours, the
rest of the ducks were picked off one by one.

Utah: just west of Provo, March 1980. We drove around the long sweep of
Utah Lake and into the valley beyond. Sagebrush lined the ground as far
as I could see. "Here is eagle country," said Joe Murphy, a professor at
Brigham Young University. "There are incredible numbers of jackrabbits, and the eagles gather here each day during the winter to feed.
Our studies show that many of the rabbits eaten by eagles have been
shot, and perhaps that is one of the reasons Bald Eagles hunt them so
well. After spending the morning hunting, if successful, they go over to
those juniper trees." Here Joe swung around and gestured toward a band
of low junipers lining the side of the valley below Mount Oquirrh.
"They are loafing or sunning perches."

   Turning in the other direction, Joe pointed out the side of an old Pony
Express station. "There's one of the valley roosts. Eagles spend the night
in those trees, particularly when the temperatures are mild. When it's
colder, the eagles go right up into a canyon. We'll see that soon." We
stopped briefly at an unimpressive group of about forty cottonwoods,
the roost, then drove on for some time before reaching a narrow canyon.
It was late afternoon; we had arrived about the same time as the eagles.
One by one they came in, very high, then dropped down to perch for
the night in conifers along the side of the canyon wall. High-pitched
cries drifted down to us. Eager to find out what was happening, I
watched closely. A high-flying eagle folded its wings and descended like
a bomb toward a limb. The occupant screamed loudly. The attacker
veered, almost caromed off, and hurtled toward another perched eagle.
The second eagle, with less time to prepare, was caught unawares and

jumped back, hastily vacating its spot. The new arrival landed heavily on the limb, then edged over a few inches to nestle in closer to a large immature next to the tree trunk. We were witnessing the nightly clamor of eagles vying for favorite perches.

Texas: northwest of San Antonio, December 1981. Jim George and his son Perry had come from San Antonio to join us on Besnard Lake one summer. Now, several months later in early winter, I was on my way with him to meet Sam Crow, a wildlife officer who was working on eagles. We had come to see the country where eagles had long been accused of preying on newborn lambs. When we met him, I asked Sam whether eagles ever actually killed lambs.

After hesitating a moment, Sam said, "Yes. As an experiment, a lamb was taken away from its mother and tethered. Some hours later it was killed by an eagle. However, that was hardly a natural situation. Usually lambs stay close to their ewe. Lambs that eagles feed on are almost invariably carrion, already dead when the eagle first arrives. It is a tough question to answer definitively. Some ranchers complain a lot. But I've noticed that those ranchers who take good care of their sheep have few losses for any reason."

It is estimated that 20,000 eagles, mostly Goldens but some Balds, were killed to save lambs between 1950 and 1970. Most were shot by hired guns who went up in airplanes. One wonders how high the price of lamb could be to justify such an expense. Or did it become more the love of a blood sport that drove people to such acts? Although eagles are no longer shot from planes (we hope), careful studies have shown that it is extremely rare for Bald Eagles to kill lambs, kids, or goats, and any negative effects of eagles on these animals are likely to be more than compensated for by the eagles' killing of rabbits, which compete with sheep for food on the range. We visited the areas where there were sheep and later saw several eagles perched in a roost along a canyon.

*W*inter is the time when eagles search for and eat whatever food they can find. It is a time when food is relatively scarce, and colder temperatures and shorter days mean that eagles have a tougher time surviving. Bald Eagles generally prefer fish (as shown in studies by B. S. Wright, in which they were given a choice),[1] but they can often be found eating rabbits or waterfowl in winter. Frequently, at this time of year, eagles gather where prey is plentiful. In areas of Utah, Colorado, and Oklahoma, Bald Eagles feed primarily on rabbits. Across the midwestern United States at wildlife refuges, Bald Eagles feed on geese and ducks. In some areas, like the salmon-spawning regions of Washington and in the waters of the Mississippi River, fish are available and are still the primary prey. Dead gizzard shad or other fish may be locally plentiful when winter ice conditions (or just a drop in water temperature for certain other fish) lead to a decrease in the available oxygen. Eagles come to pick up the dead or dying fish, often in the open water below a dam, a rapids, or waterfall, or where one river entering into another creates enough turbulence to keep the water free of ice. The poem at the beginning of the chapter describes one such location, the gorge just below Niagara Falls, where eagles were common in the early 1800s.

Eagles congregating to feed often roost communally. In cool weather, finding the most protected site and perching close together on the branches may help to reduce heat loss. The roost may also allow inexperienced birds to learn optimum hunting locations by following birds with more knowledge of the local terrain. Susan and Richard Knight, working on the Nooksack River in Washington, found that eagles tended to follow one another, with immatures (i.e., less experienced birds) more likely to follow adults than the reverse.[2] After a flood on the Nooksack River, when the usual plentiful supply of salmon carcasses was washed away, the prevalence of following increased; that is to be expected if following is more important when food is harder to find.

Roosts may also add security. The usually solitary Golden Eagle often roosts on the ground or a hillside and is vigilant and wary even when sleeping at night. In contrast, the Bald Eagle, which roosts in trees, is a sound sleeper, and communal roosting may enhance its security.[3] When numbers of eagles congregate in winter to hunt or roost, social behavior becomes much more conspicuous. Within a roost, a hierarchy develops. The oldest and most aggressive eagles usually occupy the highest perches.

Along the Skagit River in Washington, as at other wintering sites, eagles tend to leave their roosts in the morning in the hour before sunrise, often when it is still quite dark. The proportion of eagles flying was highest during the first hour of observation (7:00 to 8:00 A.M.) as eagles actively searched for food.[4] Birds that were satiated generally found a perch and stayed there.[5] When food was harder to find, the percentage of eagles seen soaring increased dramatically from 7 percent to 68 percent.[2] When the Knights set out food stations, they found that the time it took the first eagle to find the food was highly variable but on average was 35 minutes after sunrise. After the arrival of the first eagle, other eagles found the station quickly, with an average of 21 additional eagles arriving within 30 minutes of the first. Indeed, eagles home in on other eagles to such an extent that when given a choice between an unattended (i.e., no eagle present) and a nearby attended food station, the arriving eagle went to the food where there was already an eagle present in 182 out of 191 cases.[2]

Hansen, observing at similar stations in southeast Alaska, noted that there was usually a considerable wait from the time food was first discovered before an eagle started to feed, an average of 25 minutes, with early arrivals often perching nearby in a tree. Once an eagle started to feed, how-

An eagle zooms in to join others feeding on the ice (John E. Swedberg).

ever, other eagles quickly came down to the site to try to displace the feeding bird, allowing only 4.4 minutes to pass on average.[5]

When two or more eagles try to feed on the same carcass, threats, scrapping, and fighting usually occur, as the following illustrates.

> Two adults were flying in the vicinity of the Cottonwood area when one of the eagles landed near a group of crows. The eagle used a threat display as it hopped toward the crows. The crows took off leaving some carrion behind. The eagle moved over the carrion and began tearing off pieces, while the other eagle landed and sat nearby. About five minutes later, the eagle that was eating took to the air. . . . The second adult moved over the food and began eating. [The first adult returned to the ice after a brief flight.] A few minutes later, a subadult came barrelling in driving the adult from the food. The subadult landed on and mantled the food while vocalizing. The two adults moved near and after a brief stand off, . . . which included the flapping of wings and the showing of talons, there was a pause in the action. Then one of the adults ran toward the subadult driving it off the food while the other adult grabbed the food and flew away. . . . The first adult soon followed.[6]

Hansen showed that larger and hungrier eagles were more likely to be successful in competing for food.[5] Hungrier eagles show their disposition by being more aggressive and doing additional displays. A hungry eagle

Squabbles over food are commonplace on the wintering grounds (John E. Swedberg).

is more likely to approach another feeding eagle with neck extended, raising and lowering its tail, raising its wings, rapidly undulating its wing tips, and calling. Smaller eagles are less likely to steal from other eagles and more likely to catch their own food, with the smallest eagles hunting almost exclusively. Since the chances of successful theft are higher for larger and hungrier eagles, such birds choose the option to steal a fish as often as they choose to take a fish on their own. This was true even when large numbers of dead fish were easily available. Hansen argues, using game theory and firsthand observations to develop his point, that thievery is common even when food is plentiful because it is an inherited behavior that is used even when it is not essential.

Low temperatures and periodic scarcity of food mean that winter is the period of highest natural mortality. To adapt to adverse winter conditions, eagles generally employ strategies to conserve energy. In winter, birds spend proportionately less time flying than in summer. Observations along the Nooksack River in Washington suggest that flight may occupy as little as 1 percent of a day (calculated on a 24-hour basis), with eagles spending 68 percent of their time in a night roost. Of the daytime activities, fully 93 percent of the time was spent perching, 4 percent feeding or waiting to feed, and only 3 percent flying.[7] However, the proportion of time eagles spent flying and feeding in the San Luis Valley of Colorado was about twice as much, suggesting that there is variation from one wintering area to another; perhaps the higher activity in the San Luis Valley was caused in part by the wider dispersal of food.

In a study of energetics at different temperatures, Mark Stalmaster, working in his laboratory in Utah and in the field in Washington, found that an eagle needs to consume between 6 percent and 11 percent of its body weight every day—toward the lower end of the range if it feeds solely on ducks and toward the higher end if it feeds solely on fish.[8] Rabbits are a good source of intermediate value. As the weather gets colder, more food is needed. At 14°F an eagle would need to eat 11 percent of its weight in chum salmon, whereas at 68°F it would need to eat only 8 percent of its body weight. Stalmaster calculated that a ten-pound eagle would need a minimum of 13 salmon, 20 rabbits, or 32 ducks to survive ninety days at 41°F.

Immature eagles do less well than adults in time of food stress. Meticulous observations by Mark Stalmaster show that immature males, as a group, do the worst.[7] Indeed, the evaluation of food intake by eagles wintering along the Nooksack River suggested that a significant proportion of the immature males were not getting enough food to maintain their body weight, while all other groups were doing significantly better. The result is a higher casualty rate among younger eagles, perhaps particularly among young males. (Before the introduction of a winter feeding program, it was estimated that 46 percent of first-year Maine eagles died, but only 9 percent of adults.[9]) It is at such times, when the shortage of food becomes desperate, that eagles will eat almost anything. Eagles have been recorded, for example, feeding on a garbage dump on Amchitka Island, Alaska, even fighting over a morsel of bread when food was severely curtailed. It is truly a time for searching for anything edible.

Winter counts of eagles have provided a window on the North American population. For many years, an annual January count has been made throughout the United States and parts of Canada, coordinated by the National Wildlife Federation. About 14,000 eagles were seen in 1982.[10] Because there is not uniform coverage, that figure should not be taken to represent the total of eagles in the contiguous United States. Indeed, an evaluation of Christmas bird counts suggests that the real number of Bald Eagles wintering in the contiguous United States is closer to 20,000.[11] The percentage of immatures on the more recent counts has been between 33 and 41 percent, a more reasonable proportion than in the DDT days, when

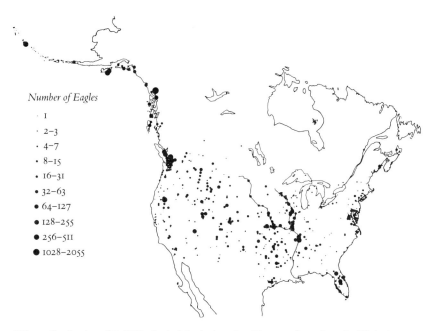

Winter distribution of Bald Eagles in North America. Data are from the 1985 Christmas bird counts (American Birds *Volume 40, 1986, pages 606–1032*).

it was 22 to 27 percent across the United States and less than 20 percent in some areas.[10] The winter censuses also show that eagles are most likely to be found in locales where the mean temperature is near freezing (Appendix 9). Immatures, however, are relatively more plentiful farther south (Appendices 8 and 10).

Many studies have been done on wintering eagles, covering a wide range of habitats and locations. It would be impossible to cover all of them, but we shall discuss several illustrative examples. Five of the sites covered, those in Utah, Colorado, Oklahoma, Texas, and Missouri, fall within the primary winter range of eagles from Besnard Lake. (All have at one time or another been visited by J.G.)

Chilkat River, Southeast Alaska

Coming down from the mountains of the northwestern corner of British Columbia, a river emerges to cross a flat valley in the Alaska panhandle before emptying into the Pacific Ocean. Each fall, thousands of salmon return from the ocean and come up this river, the Chilkat, to spawn in the shallow gravel beds. Here, near Haines, Alaska, a huge gathering of 3,000 or more eagles is lured by an easy meal—the salmon die after they have spawned. When the weather turns cold and salmon carcasses freeze, the birds either roost in the black cottonwoods along the river and wait for warmer weather or leave the area.[12] Gradually, as fewer and fewer salmon are available, the eagles disperse along the Pacific Coast, with some possibly going as far south as Washington or Oregon for the remainder of the winter.

Skagit and Nooksack Rivers, Washington

In northwest Washington State, just below the Canadian border, there are two excellent salmon-spawning rivers, the Nooksack and the Skagit. Originating in the Cascades, these rivers descend swiftly through largely coniferous forests. In the lower reaches there are extensive gravel beds; here in late fall and winter, eagles come to feed on spawning salmon. Although there are several species of salmon in these waters, the chum, which spawn from November to January, are the most important to the eagles. The timing of chum spawning coincides with the needs of the eagles moving southward along the west coast in late fall. The chum spawn in side channels and in shallow areas of the main channel. After spawning, the carcasses of the fish wash up on the gravel bars and islands where the eagles come to feed. In some years, as in the winter of 1979–80, floods may wash away the chum carcasses; then the coho salmon in the tributaries become important to eagles.[4] The number of eagles near the rivers in the winter reflects in part the availability of food. Some years there is a much better spawning run than others. For example, in the winter of 1978–79, a good year for chum, there were estimated to be about 300 eagles along the Skagit River. In 1979–80, a poor year for chum, there were only about 100 eagles. The pattern of eagle use and salmon spawning on the Nooksack was the reverse; 283 eagles were found in a stretch of the Nooksack in 1979–80, when the chum salmon run there was unusually good, but there were only about 100 birds in 1978–79. The findings suggest that eagles move back and forth in this region from one spawning river to another to take advantage of the local conditions. Radiotelemetry studies have confirmed such movement.

During the day, the eagles along the Nooksack and Skagit tend to perch in deciduous trees, particularly red alder and black cottonwood, along the river; the taller cottonwood is preferred.[13] At night, roosting occurs both in the deciduous trees along the rivers and also in the surrounding coniferous forests. Along the Nooksack, eagles clearly preferred roosts situated in dense stands of live Douglas-fir and western red cedar; they may provide a more favorable (warmer) microenvironment in the winter than deciduous trees. Although there are some communal roosts along the Skagit River, studies using radiotelemetry have shown that many of the eagles roost singly and that they tend to choose a different spot each night.

Rush and Cedar Valleys, Utah

These two intermountain valleys lie on the west side of Utah Lake. Cedar Valley, the easternmost, lies between the lake and a chain of mountains including Mount Oquirrh on the north side and the East Tintic Mountains on the south side. Rush Valley, the next valley to the west, is reached by passing between Mount Oquirrh and the East Tintic Mountains. On its west side are the Onaqui Mountains. The lowland flats that form the basins of these valleys were the sediments from the huge, prehistoric Lake Bonneville. Bald Eagles were first noted to use the valleys in 1960 and have been studied extensively since then by Joe Murphy and his graduate students.[14] Bald Eagles arrive in late November or December and stay until March, with immatures occasionally staying until the first few days of April. Two mountain roosts, the first in the East Tintic Mountains and the second near Ophir on Mount Oquirrh, are the night resting spots for most of the eagles that visit the valleys. In midwinter, two valley roosts in cottonwood trees also frequently have eagles. As an example of how these eagles can be distributed, on February 6, 1969, there were 47 eagles in the Tintic roost, 61 eagles in the Oquirrh roost, 11 in the Vernon valley roost, and 10 at the Fairfield valley roost.

Each morning, the eagles generally leave their roosts early to go to selected sites in the valleys, where they encounter their primary prey, blacktailed jackrabbits. The Bald Eagles often hunt in small groups with one eagle flushing the prey for another to catch (see Chapter 4).

After a successful hunt, the eagles retire to loafing perches in the low juniper and piñon pine trees that line the sides of the valleys. Toward dusk the eagles return to their mountain roosts. The roost sites are located in bowl-shaped confines or canyons in the mountains that are well protected from the wind; they are about 6,000 feet above sea level, or about 1,200 feet above the valley floor. The eagles usually perch in Douglas-firs on a ridge within, but 300 feet above the base of, the canyon. As they come into the roost, there is much competition for favored sites. Eagles arriving late will try to displace a bird that came in earlier. Loud protests erupt at intervals from the perching birds whenever a new arrival threatens to intrude. So close are the eagles perched at times that a new arrival may knock more than one eagle off the desired limb. By dusk the eagles may be so close together as to be touching one another.

San Luis Valley, Colorado

In the south central part of Colorado there is a huge, high intermountain valley as large as the state of Delaware. To the west rise the peaks of the San Juan Mountains. To the east is the Sangre de Cristo range of the Rockies. In winter it is a cool, semidesert region. Black greasewood and inland saltgrass carpet the northern parts of the valley. In the southern parts there is extensive rabbitbrush, except for areas where farmers with their circular, central pivot irrigation systems have taken over. The valley is the winter home for large numbers of Bald Eagles. Estimates of 250 in 1976 and 170 in 1981 give some idea of the size of the population.

Beginning in about 1976, a graduate student, Al Harmata, spent long hours studying the eagles of this valley to provide some perspective on their ecology.[3] Each winter the eagles arrive in late November or December. Each bird settles into a home range of 23 to 309 square miles, much larger than that used in summer. At night the eagles usually roost communally, preferring tree stands near water (usually one of the major rivers, the Rio Grande or the Conejos), and as far as possible from a county road or an urban center. Most roost sites in the valleys are in cottonwoods; a few are in ponderosa pine on mountain sides. Between six and seven each morning, the birds leave their roosts to go to a hunting perch. As the day progresses, an eagle will move from one perch to another, staying from minutes to hours on each. In the middle of the day, when thermals are best, eagles may spend quite long periods (two or more hours at a time) hunting on the wing. Such flights are often quite localized in the amount of territory covered (0.8 to 5.8 square miles).

Six of ten adult eagles that Al Harmata caught, equipped with radiotransmitters, and then followed long enough were mated on their wintering grounds. Mated birds generally, but not always, had a smaller winter range than unmated ones, perhaps because the pair was able to occupy better habitat.

Hunting ranges usually include a section of river, but most of the area covered is upland, where rabbits are preyed upon, some distance from the river. Eagles sometimes defend particular perches, but they make no attempt to exclude other eagles from their winter range; therefore the ranges of different birds often overlap. The aggressive encounters that occur are mostly over food.

In January, the eagles in the San Luis Valley feed on a roughly equal mixture of mammals (mostly white-tailed jackrabbits) and birds (usually ducks, with Mallard being most common). From February to March the proportion of ducks decreases, the proportion of rabbits increases, and as the river opens up the eagles start feeding on fish.

Oklahoma

The rivers and lakes of eastern Oklahoma, with their forested borders, are home to most of the state's wintering eagles. The western half of the state is flatter, has fewer lakes and trees, and, with the exception of Great Salt Plains Lake and the adjacent Salt Fork of the Arkansas River, has relatively few wintering eagles. Most of the lakes of the state were created by dams built starting in 1921. Bald Eagles feed predominantly on fish and geese; small groups of eagles in upland areas that hunt rabbits or search for carrion are the exception. An annual count by Jim Lish suggests a state total of 600 to 800 birds; however, the count is far from complete; although the major roosts are covered, only a small proportion of the total shoreline is censused, and eagles on upland areas are not systematically or randomly sampled.[15] The distribution of eagles varies with the severity of the winter — more eagles inhabit the northern half of the state if the winter is mild, but if the winter is severe and the northern lakes freeze over more eagles can be found in the southern half.

Texas

Texas is the southernmost region of North America where considerable numbers of Bald Eagles winter. Occasionally, migrant Bald Eagles do wander into Mexico (in 1977 two pairs even nested successfully in Baja California), but present evidence suggests such birds are few. It is typical of migrating Bald Eagles that the immatures move south first and go farthest south; the result is that most eagles wintering in Texas are less than four years old. Many of these birds intermix with Golden Eagles, hunting upland regions and feeding primarily on carrion or rabbits. Accurate censusing of such a winter population is difficult, and the total of 397 Bald Eagles found in the winter of 1979 is probably far below the real number present. Some sort of random sampling survey will be necessary to provide a more accurate estimate.

Squaw Creek National Wildlife Refuge, Missouri, and the Mid-American Plains

From the Rockies to the Mississippi River, from the Canadian border to northern Texas, the story of Bald Eagles in winter is closely tied to the stories of the wildlife refuges and the lakes created as the U.S. Army Corps of Engineers built reservoir after reservoir. One such area is Squaw Creek National Wildlife Refuge in the northwestern corner of Missouri, a refuge nestled just inside the loess hills along the eastern side of the Missouri River. Here, a series of ponds is home to tens of thousands, sometimes hundreds of thousands, of migrating geese and up to several hundred eagles each fall and winter. Squaw Creek, a narrow trickle of a stream in summer, drains the nearby hills. It has been dammed and channeled so that water levels on the refuge ponds can be lowered or raised. The water level is fluctuated to keep down the growth of cattails, which threaten to take over the marsh, and to enhance the growth of plants that provide food to be ready for the fall, when the sky is dark with migrating vees of geese. Bald Eagles, mostly immatures to begin with, usually start to arrive in early October. As fall progresses, the concentration of eagles grows and may reach several hundred individuals by late November and December. In some years the refuge ponds freeze early and the geese and eagles are pushed farther south to other refuges. Occasionally, the ponds don't freeze at all; eagles and geese stay the winter long. The eagles appear to feed primarily on carp in the refuge pools and on crippled geese and ducks. At night, most eagles retreat from the pond to a sheltered roost in trees at the north end of the refuge.

Curt Griffin has studied in depth the home ranges of eagles at the Swan Lake National Wildlife Refuge, Missouri. This refuge is similar to Squaw Creek but about 120 miles southeast. (A marked eagle from Besnard Lake visited both Squaw Creek and Swan Lake one winter, showing that eagles move from one to the other.) Eagles at Swan Lake had winter ranges of 1.3 to 26.6 square miles, much smaller than those in Colorado but still larger for the most part than summer breeding ranges. The size of the home ranges used at Swan Lake varied with the availability of food. In 1975–76, when waterfowl were less available, the mean winter range of eagles was 18.5 square miles. In 1977–78, when waterfowl were abundant on the refuge, the mean winter range decreased to 7.1 square miles.[16]

Mississippi River

Along the Mississippi River, observations of wintering Bald Eagles in the fifties had a particular significance. Here Elton Fawks, a bird-watcher, alerted the country that serious problems with Bald Eagle reproduction had extended beyond the east coast and reached the waters of the Mississippi. Fawks, a descendant of the family of David Crockett the frontiersman on one side and the family of John L. Sullivan the heavyweight champion on the other, came when he was five years old to East Moline, Illinois, on the "Father of Waters." At the age of ten he became interested in birds. In 1948 Elton was elected president of the Tri-City Bird Club. A visit from Charles Broley in the fall of 1954 stimulated him to look closer at the eagles wintering along this stretch of the Mississippi.[17] By the early fifties, the number of eagles wintering near East Moline was averaging around 80, with a reasonable balance between the brown-headed immatures (30 to 40 percent of the total) and adults. By 1958, however, Fawks noted an alarming drop in the percentage of immature Bald Eagles. One count during that winter showed 58 adults with only 1 immature. A second count had 58 adults and no immatures. The lack of immature eagles suggested that eagles on their breeding grounds to the north were having difficulty producing young. The winter of 1958–59, Elton made numerous counts in the region near East Moline. From all counts, made on many different days, the total was 590 Bald Eagles, with 103 (17.5 percent) immatures. In 1959–60, the total of similar counts was 1,123 eagles with 204 (15.4 percent) immatures. The counts were on a variety of dates and did not necessarily cover the same ground; however, the total number of eagles seen was impressive, and the percentage of immatures found was only about half of what it should have been. Together with Charles Broley's alarming findings in Florida (see Chapter 1), these observations led to the first attempt at a nationwide eagle survey. The National Audubon Society became interested, and in 1960–61 Sandy Sprunt coordinated the effort. As part of this effort, Elton reported 397 eagles with 42 (10.6 percent) immatures along his stretch of the Mississippi River on February 9, 1961. The national count confirmed

that the overall percentage of immatures was very low, and it was an important factor in alerting people all over the continent to the imminent danger
facing the Bald Eagle.

Today, each winter large numbers of Bald Eagles congregate at good
feeding spots along the Mississippi River between Minneapolis and St.
Louis. Counts along that section of the river have shown an increase in
the proportion of immatures, which was up to about 30 percent in 1980.[17]
Prescott, Wisconsin, about 25 miles south of Minneapolis where the St.
Croix River enters the Mississippi, is usually the northernmost gathering
place. Here, the turbulence from the merging of the two big rivers always
leaves some open water where eagles can fish. As I frequently observed
during the winters I lived in Minneapolis, the eagles feed largely on gizzard shad, which they scoop up from the water surface or pick up along
the edge of the river by walking on the ice or into the water. Though there
may be eagles that winter farther north, usually a cover of ice and poor
hunting conditions are limiting. The location of the peak number of wintering eagles appears to vary from year to year, depending on local food
availability and ice conditions. Keokuk, in southern Iowa, is certainly one
of the hot spots. At some sites, for example near Eagle Valley, Illinois,
eagles have been found feeding on pig entrails thrown out as refuse by a
farmer.

The eagles usually roost in well-sheltered valleys, just back from the
river, but may move to a more sheltered location if the wind shifts on a
cold night. South of Reelfoot Lake, just east of the Mississippi in Tennessee, there are few eagles at any single site, though individual birds may be
found all the way to the Gulf of Mexico. All told, the river itself and adjacent areas may have up to 4,500 eagles in midwinter.[11]

Maine and Nova Scotia

In Nova Scotia and Maine, many of the breeding eagles stay for the winter. In Nova Scotia, the primary nesting grounds are on Cape Breton Island,
but the main wintering regions are farther south along the Shubenacadie
River and in the Annapolis Valley.[18] Some eagles also frequent the icefront
in Northumberland Strait and the Gulf of St. Lawrence between Prince
Edward Island and eastern Nova Scotia, where gray seals whelp, leaving
behind their placentas and the occasional dead pup for the eagles. A few
eagles are also found around the Atlantic coastline in the southern parts of
the province where there are wintering seabirds. For some years P. C. Smith
and his colleagues have been involved in studies at feeding stations either
along the Shubenacadie River or at Gaspereau in the Annapolis Valley.
The station at Gaspereau is of particular interest since it has been in existence since 1963, with a considerable increase in the number of eagles using
it, from 3 in 1963 to 52 in 1980.[19] Using blinds and telescopes, Smith has

been able to read the bands of some of the birds. He found that many of them arrive from northern Nova Scotia, and one was from a nest on Prince Edward Island.

Similar feeding stations have been set up along the coast of Maine, where Bald Eagles were severely affected by DDT.[9] Winter feeding was deemed to be critical to provide these birds with a "clean" diet (see Chapter 12). The program, by assisting eagles during the stressful winter period, also appears to have increased the survival rate of eagles during their first year from 54 percent to 73 percent. In addition to local birds, eagles from Nova Scotia have been found to winter there, as well as a 2½ year old that came all the way from Besnard Lake, Saskatchewan.

12. *Stewards or Destroyers?*

Ill fares the land, to hastening ills a prey,
Where wealth accumulates, along with spray.
Chemists and farmers flourish at their peril:
The bird of freedom, thanks to them, is sterile.
E. B. White

Gary—

Toronto, Ontario: October 1985. The crisp autumn air was too much to resist. After weeks of not having a day off, I needed some time to get away, get outdoors, and go birding. How ironic that I chose a profession out of love of nature, yet most of my days were spent behind a desk in Canada's largest city. Today, however, I would join the flocks of migrant birds and take refuge on the Toronto Islands.

After just a short ferry ride, the world around me changed dramatically. Somehow the air smelled sweeter and the sky looked bluer. The looming city skyscrapers that crowded the waterfront were no competition for the cottonwoods and willows; the huge imposing trunks that dotted the parkland dominated the island skyline. My wife, Heather, and I were rejuvenated by the cold Lake Ontario wind and the crashing of the waves on the breakwater. Making our way to the small corner of land reserved as the wildlife sanctuary, we found pleasure in watching the Canada Geese graze on the manicured lawns. The tangled mats of grapevine, dogwoods, and willows that comprised the sanctuary were a haven for many birds. Today though, we rummaged through the thickets, craning our necks every which way in search of Saw-whet Owls. Finding the birds was unimportant; the day's success would be measured by how many colored leaves we saw and how much fresh air we breathed, and likely by how good it felt to be physically, instead of mentally, tired.

As I peered through the bushes, a large dark bird caught my eye as it passed just over the treetops. In an instant it was gone. I didn't see any definitive field marks, and I didn't see it for long, for the brush was too thick, but I knew it was a young Bald Eagle. I shouted to Heather, but to no avail. When I told her what I thought I had seen, she just raised her eyebrows as if to say, "Come on now, Gary. This is Toronto." It was what I would have done if the situation were reversed. Could I be wrong after all those years of virtually living with eagles? It had been three years since I saw my last eagle, having been committed to my thesis and other research in the meantime. Did I really long to see an eagle so badly that I had just imagined seeing one? Common sense and my nagging scientific training told me I was probably mistaken, but damn it, that bird really *felt* like a Bald Eagle. It had, even in that short glimpse, a certain imposing presence that just can't be matched in the bird world.

My heart was no longer in the owl hunt. For the rest of the day my eyes were glued skyward. I still had the bird on my mind when we reached the ferry dock to go home. All of a sudden Heather shouted and pointed above us, "There's your bird!" The size, shape, and all-brown plumage with white underwing coverts were unmistakable — a juvenile Bald Eagle. It passed low over our heads and circled slowly, as if to give us a really good look, then soared away across the harbor. My first thought was that the bird had probably been one of those released in southern Ontario as part of an effort to reintroduce the species to its former haunts. However, the bird carried no wing tags or other markers; it was probably a wild eagle untouched by humans. Somehow, that made the sighting all the more precious.

A passing Herring Gull gave the big bird a halfhearted swoop, and an Oldsquaw or two rose from the water. In no time the eagle was above the Toronto lakeshore and soaring above a traffic-clogged expressway. With the compressed perspective of the binoculars, the scene of the eagle against a background of concrete, glass, and steel was transformed into a surrealistic painting. This was not how I was accustomed to watching eagles. Heather never imagined that her first sighting of a Bald Eagle would be anything like this, and frankly she was a bit disappointed. The eagle soared on for about half a mile before cruising back over the harbor to the islands. We lost sight of the bird as it flew over the treetops of the nature sanctuary.

While boarding the ferry homeward, I chatted with a young bird-watcher. He too had seen this, or a similar eagle, three weeks ago. Perhaps the bird had been on the islands all that time. I was concerned. Pesticides had delivered the coup de grace to eagles breeding on Lake Ontario's shores just thirty years ago. Would this bird eat a fish contaminated with one of the scores of chemicals introduced in the post-DDT era? Would it catch an ailing duck whose tissues were loaded with lead and mercury? The best I could hope for was that this bird would move on and find a healthier place to live. I wondered if the lake would ever be clean enough to welcome the return of the eagle.

*T*he Bald Eagle is pervasive, symbolically, in North American culture; its name or caricature adorns currency and represents hundreds of commercial enterprises and their products. In Ontario alone there are nineteen lakes and some forty place-names bearing the name "Eagle." As the Golden Eagle is rare in the province, surely most locales were named after the Bald Eagle. Millions of people derive pleasure from seeing eagles in art and film if not the wild.

The bird's best-known role is that of emblem of the United States. No doubt, it was the association of eagles with power that attracted the founding fathers to the species. I prefer the analogy of Stephen Fretwell, then professor at Kansas State University. In his article "Interview With a Bald Eagle," he relates the events of a meeting between a reporter and an eagle.[1] When asked, "Wouldn't you be a better symbol for us if you hunted live prey, and didn't feed on carrion?" the eagle replies:

> Actually, the fellows who made me your symbol were more right than they knew. . . . A fish dies and is washed up on shore. It looks bad and smells worse, is good for nothing, despised by all. I come and eat it and turn that fish (if you'll forgive my immodesty) into a soaring wonder, a majestic greatness that stirs the heart of creatures everywhere, including men. Isn't that true for America, too? America was built by religious rejects, crooks, and poor people. All human waste, all looking bad and smelling worse, despised by all. This country consumed these people and . . . made them into a nation. . . . Like me, you . . . took the waste of the world and made something wonderful.

In retrospect it is difficult for many of us to conceive of how people of this continent could have allowed, let alone participated in, the needless, often wanton, destruction of such a marvelous bird. By 1967, so few eagles remained in many of the locations where once they had flourished that the Bald Eagle was officially declared endangered throughout much of the United States, later in parts of Canada. In this chapter we look at the current relationship between humans and Bald Eagles. We have come a long way toward appreciating the Bald Eagle, but there are still many ways in which we could be better stewards in our dealings with eagles and the environment that we share.

Mortality

Populations may suffer from factors directly affecting the survival of individuals or from the more subtle, indirect effects that lower reproductive success and hence the capacity to overcome attrition due to natural mortality (mostly starvation and disease). For long-lived birds such as eagles, direct mortality is of greater consequence to the continued existence of a population than is a depression in reproduction.[2]

The most serious of the direct human-induced mortality factors is shooting. Typically, between 20 percent and 60 percent of all eagles found dead have been shot.[3] The problem may not be as widespread as it once was; however, in an endangered population the death of even one bird can be a great loss. In 1980, the male of New York State's last natural, wild breeding pair of Bald Eagles was shot.[4]

As is typical of many raptors, it is the young birds that are most susceptible to being shot. It has been postulated that hunters confuse the brown, immature eagles with hawks (not that this lessens the crime), but the inherent tameness of young eagles no doubt makes them an easier target as well.

Although the thoughtless gunner may shoot eagles for sport, another class of criminal, the poacher, shoots eagles for profit. There is a lucrative illegal market for wildlife products. Near one wildlife refuge in South Dakota alone, between 200 and 300 Bald Eagles were killed over a three-year period ending in 1983. Most were caught in baited traps or shot at night roosts. Feathers, bones, talons, and bills were used to make "authentic" reproductions of American Indian artifacts (warbonnets, peyote fans, and so on) for sale in America and Europe.[5]

Not all human-related mortality is as intentional as shooting. Each year eagles die from a myriad of causes connected with human alteration of the natural landscape: electrocution, collision with power lines or motor vehicles, entanglement in fishing line. Some problems may be locally more important than others. One of the more serious, widespread causes of mortality is the accidental trapping of eagles in leghold traps set for other animals.[6] There are actually two distinct trap-related problems with different kinds of solutions.

First of all, eagles often die in traps or by eating poison set out for other predatory animals, such as wolves, coyotes, foxes, bobcats. Sometimes the killing is done in the name of predator control, but more often it is incidental to the provision of furs for the fashion industry. Eagles, being carrion eaters, are attracted to the site by bait. The use of poison cannot be justified for it indiscriminately kills whatever animal eats the bait, and sometimes the unintended victims far outnumber the target species. Trapping of eagles can be effectively eliminated by placing the bait at a distance from the trap or by using scents. Some states are now banning or restricting the use of baited sets because of the number of eagles caught. When Minnesota did so in 1980, the number of raptors with trapping injuries admitted to the Minnesota Raptor Research and Rehabilitation Program fell dramatically.[7]

The second kind of trapping problem is one that Bald Eagles, more so than most raptors, fall victim to frequently. Although no bait is used, traps set for muskrats and some other furbearers catch eagles because they are set on top of mounds where eagles perch. The problem can be effectively circumvented by using underwater sets, such as the Conibear trap.

It is difficult to get accurate estimates of the number of birds being caught; trappers often live in remote locales and don't like to admit to catching eagles. However, there is ample evidence to suggest that trapping can be a serious cause of mortality. Twenty-one Nevada trappers caught 109 eagles (mostly Goldens) in just four months.[8] One trapper in Saskatchewan told me that one spring he caught eighteen Bald Eagles. On several occasions at Besnard Lake we have seen or heard of eagles flying around with traps dangling from their legs. Most of these sightings were around spawning streams in spring where food can be obtained easily.

In 1982 we came across what appeared to be a survivor of a trapping incident. An eagle perched on the rocky shore of Besnard Lake flushed as we boated past it; however, it immediately foundered and fell into the water. The bird was easy enough to catch. It was a female and almost in adult plumage. She had just eaten but was so emaciated that she could not fly given the weight of food in her crop. Her keel, or breastbone, jutted out sharply, for she had little breast muscle, a sure sign of a bird in poor condition. That was not a surprise, for all the toes on one foot were missing. Having few, if any, alternatives, we put a band on her good leg and set her on shore for the night. We were surprised the next morning to find her gone. We were even more surprised five days later to see her flying about a few miles from the release site. Unfortunately, this bird's freedom would likely be limited. Although the injury had healed well, the other foot was in bad shape.[9] Having to bear the entire weight of the bird had taken its toll on the remaining leg; the talons had grown curled, and the scales on the underside of the foot were worn raw in places.

Contaminants

Chemical contamination of our environment is pervasive; innumerable insecticides, herbicides, and industrial waste products pollute our air and waterways. Many contaminant problems for wildlife are local—selenium in California—and others are widespread, even global, in nature—DDT and PCBs, for example. There are three major means by which a pollutant can harm a Bald Eagle population: (1) direct mortality—birds are killed through direct toxicity, (2) sublethal effects—birds become more susceptible to death by other factors, or reproduction may be impaired, and (3) food supply is diminished—birds can no longer find enough prey to survive or reproduce.

The cause-and-effect relationship between a contaminant and a decline in eagle numbers is difficult to prove in the field. The pollution level in the environment may be such that it would not appear to be great enough to be a cause for concern. However, because of biomagnification—the accumulation and concentration of a pollutant as it moves up the food chain—predators such as eagles can be exposed to large doses in a single meal. It is easy to underestimate the effect that a contaminant is having on a population. When contaminants are toxic, their effects are rarely immediate, and so the bird may die away from the contaminated site. The carcass of an eagle that has died from pollution poisoning is also less likely to be found than that of a bird shot or trapped.

One recent cause for concern has been the toxic effects of lead; more than 120 deaths of Bald Eagles from lead poisoning have been reported in the last twenty years. Eagles ingest lead shot from spent shotgun shells contained in their prey. Waterfowl shot but only crippled by hunters frequently fall victim to eagles in winter. Substantial numbers of waterfowl across the continent carry lead shot within them. For example, at the Swan Lake National Wildlife Refuge in Missouri, where some Besnard Lake eagles winter, 43 percent of 20,759 adults and 12 percent of 9,945 immature Canada Geese contained at least some shot.[10] To compound the problem, waterfowl accidentally ingest shot, which not only leads to their demise (it is estimated that approximately two million waterfowl die annually this way) but also provides another means whereby eagles can be poisoned. Although eagles often regurgitate the shot along with some of their prey's feathers and bones, some lead enters the gastrointestinal tract.[10]

So serious has the problem of lead poisoning become that legislation, specifically to protect the Bald Eagle, has been passed to require that steel shot be used in some areas where eagles winter. The National Wildlife Federation believed that the U.S. Fish and Wildlife Service had delayed too long in protecting the Bald Eagle, and so in 1985 the federation filed suit, requesting an injunction to enforce the Endangered Species Act by banning lead shot in the 22 counties of the five states that had refused to establish nontoxic shot zones. The federal district judge favored the National Wildlife Federation and issued the injunction. The target date for the total exclusion of lead shot in the lower 48 states is 1991. The problem is under study in Canada.

Other pollution issues are rarely as easy to solve or to mitigate as the lead shot problem. The time lag between identifying a problem and passing protective legislation can be long. Much damage had been done by DDT before its use was finally curtailed. Although there is evidence that several eagle populations are recovering from the DDT era, now is no time to be complacent. Even putting aside the potential deleterious effects of the thousands of new chemicals in use, DDT may still be a threat. Some currently used pesticides contain quantities of DDT. In California alone, some 40,000 pounds of DDT are to be released into the environment over three years as just one brand of pesticide is phased out.[11]

Acid rain, although not a contaminant in the sense that a pesticide is, is perhaps the most serious current environmental problem in North America, and in many other parts of the industrialized world. There are many potentially deleterious effects of acid rain. Acidification can lead to an increase in heavy metal contaminants in fish, which in turn may be toxic to eagles. Of greater concern, however, is that acid rain can reduce fish

Nesting Bald Eagles disappeared from many areas as a result of pesticide contamination (Gary R. Bortolotti).

populations. Game fish are often the first to disappear as a lake becomes more acidic. Some common prey species of eagles, such as white suckers, are more tolerant, and their populations may even increase in the short term. However, if acidification continues, a lake's fisheries can be exterminated.[12] It is unlikely that many Bald Eagle populations could survive without fish as a prey base. Acidification of our environment can be stopped, but government action to force industry to install pollution control devices has been slow, and lakes continue to die.

Eagles have proven to be valuable biomonitors of environmental quality. The eagles that we seek to preserve for their majesty, grace, and beauty may repay us manyfold by serving as an early warning system for environmental hazards. Is a world that is not clean enough for Bald Eagles clean enough for us?

Habitat Loss and Management

The degradation of habitat, be it by chemical contamination, resource extraction (logging, mining), or human encroachment, is probably the most serious problem facing wildlife today. It is a complex issue because of the multitude of requirements for each species and the multitude of human interests that must be considered in habitat management. A good example of how the needs of both wildlife and people can be accommodated is the case of the Bald Eagle in Washington. In 1984 the Washington State Legislature passed an eagle protection act that authorized the Department of Game to establish and enforce protection of areas around Bald Eagle nests on private property. Landowners erupted in a furor. The department hired a professional mediator who brought together representatives from the local Audubon Society, Indian tribes, forestry industry, and cattle and dairy men. For ten months they met and developed regulations for protection of critical Bald Eagle habitat. Historic enemies got together and today take pride in the final product.

On the breeding grounds, management of the Bald Eagle primarily involves protecting nest trees.[13] Eagles usually nest in old-growth forests, the same ones that are much sought after by the lumber industry. Where nest trees are on private lands, time needs to be spent explaining to the owners, who may even be unaware that eagles are nesting on their properties, what it is that eagles require for breeding success. Management plans generally involve establishing a buffer zone around the nest tree, inside which logging and other human activities are totally or seasonally restricted (Appendix II). Wise management also includes provisions for the future; forestry guidelines should stipulate that a certain percentage of trees be left standing after logging to ensure that there will be nest trees in the future.

Much of eagle country in pristine areas is not being actively managed for eagles. Habitat in northern Canada, for example, has been protected until recently by virtue of its remoteness. However, in the last decade in Saskatchewan, there has been an explosion of roads into what were wilderness areas. Initially they were built for resource extraction, especially pulp and uranium. Roads open up remote areas to cottage developments and recreational facilities such as fishing lodges, thus endangering nesting habitat; both people and eagles like to build their summer homes on shorelines. As the human population increases, so does the potential for human disturbance. Lakes near Besnard that have had a developed shoreline and access road since the mid-fifties support few nesting eagles, although they probably did historically. As yet we have no proof of a cause-and-effect relationship, but such lakes may have poor prey populations because of excessive exploitation by sport and commercial fisheries, which in turn has hurt the eagles. Comprehensive management planning is necessary to consider food requirements of nesting eagles and potential conflicts between eagles and fishermen.

At an important stopover point for migrating Bald Eagles, Glacier National Park, Montana, the problem of overfishing is at the forefront of management concerns. Here sportfishing for kokanee salmon may be at odds with what would be in the best interest of the eagles.[14] The problem of how to divide the salmon between humans and birds is not easily solved.

Habitats along migratory routes and wintering grounds are as important as, if not more important than, summer habitats. It is thought that Bald Eagle populations are limited by winter conditions, that is, the factors responsible for most natural mortality.[15] Although protecting the food supply is a critical component in managing the winter habitat, less obvious considerations, such as protecting communal night roosts from logging, are also important.

Not all current news regarding eagle habitat is bad. Dams built to generate power or control flooding have created new winter habitat. The establishment of wildlife refuges and reserves has also helped. Preserving large tracts of habitat is obviously beneficial to the species, but many significant management projects have been on a small scale and local in scope. The Bald Eagle in Arizona is a good example of a breeding population facing several problems that are not relevant in other parts of the bird's range. In the area's hot, dry climate, eaglets in cliff nests may die from heat stress, and so wildlife managers have experimented with erecting shading devices.[16] Although tree nesters do not have this problem, there are few suitable trees for the eagles to build in. The large riparian trees used for nesting have not been reproducing well. Because large stretches of the banks of the lower Verde River are devoid of trees, there are also no perches for the eagles to fish from. Therefore, poles about fifty feet tall with cross-arms were erected as a short-term solution until trees from plantings or natural regeneration can take over.[17]

For some raptors, such as the Osprey and Ferruginous Hawk, provision of artificial nest platforms can be an effective means of increasing the density of the breeding population. However, it is difficult to encourage eagles to nest on man-made structures. It may take years for a tower or platform to be accepted by a pair unless it is a reconstruction of a nest that has fallen down. Artificial nest structures are unlikely to be of much value anyway, since most Bald Eagle populations are probably not limited by the availability of nest sites. But provision of nesting platforms may prove useful where fires or logging operations have removed the big trees, or in arid parts of the West where trees are few and far between.[18]

For some populations, problems on the breeding grounds can be solved by working on the winter habitat. A good case in point is the Bald Eagle in Maine. In the sixties and seventies the productivity of this population was poor because of high levels of pesticides in the food chain. Until the environment flushed itself of DDT, the short-term solution was a winter feeding program. Bald Eagles were provided with a source of "clean" food in an otherwise contaminated environment, with the hope that pesticide levels would be kept low enough so that the birds could reproduce successfully. Between 1981 and 1985, more than 216,000 pounds of carrion were distributed among four major wintering areas, and now Maine's Bald Eagles are on the rebound.[19]

Human Disturbance

Disturbance is not a simple issue. Human interference with eagle behavior can take many forms and ultimately may have varied results. During the breeding season, the presence of humans near an active nest can cause the parents to abandon their eggs or tiny chicks, or to be kept off the nest so long that the young die of exposure.

Much of what is generally considered to be disturbance is probably unintentional—people casually fishing or hiking too close to a nest, for example. It is difficult to predict what kinds of human activities will harm individual birds or the productivity of populations. Whether or not an eagle will react to humans may depend on how far away the people are, what they are doing, and whether the disturbance is moving toward, parallel to, or away from the nest, to name a few possibilities. Similarly, whether or not there will be a detrimental effect may depend on several factors, including the stage of nesting cycle, weather conditions, and the duration of the disturbance.

Studies of the effects of human activities near eagle nests have yielded variable results, although most suggest that people have a negative effect on nesting success.[20] It is easy to be misled or to get false impressions of the influence that humans have on the productivity of nesting eagles because of examples of a few tame individuals. There are instances of eagles nesting close to humans with little ill effect; in 1985, one pair of Bald Eagles in Michigan even nested in the median strip of a four-lane highway.[21] However, that is not typical behavior.

Some eagles are fairly tame, but others are extremely upset by the presence of humans even hundreds of yards away from their nest. Much of the variability in behavior may perhaps be attributed to learning. Experience with a specific kind of disturbance generally has one of two effects on an eagle's subsequent behavior toward that disturbance. The bird may habituate to it, that is, show no adverse reaction, for it has learned that there is nothing to fear. Alternatively, eagles may become so sensitized that they react with ever-increasing intensity. We have seen good examples of both kinds of responses at Besnard Lake.

When our eagle project began in 1968 only a few fly-in fishermen and trappers used Besnard Lake. It was tough getting close to an eagle then, for they generally flushed far in advance of the boat. Today the birds are considerably tamer, and boats often pass close to them without eliciting much reaction. However, pairs that were repeatedly disturbed by my weekly measurements of the eaglets became highly sensitive to the intrusions. Interestingly, though, an overt reaction was elicited only by us and not by recreationists. Even though our boat was the same model as the

A treetop blind high in a white spruce (Gary R. Bortolotti).

one used by a fishing camp and by many other visitors, the eagles could definitely recognize something about us. Many times we saw an adult perched calmly near its nest while one or two fishing boats anchored a few hundred yards away. However, when the bird first caught sight of our boat more than half a mile away (much farther than the other boats), the adult would take off, fly over to us, and circle overhead calling excitedly. Other researchers have told us about similar experiences. We were quite fortunate that these responses were so selective, for if the eagles had reacted strongly to all humans the disturbance may have been excessive. Fortunately, we had no deleterious effect on the success of individual pairs or the population.[22]

We have actually used the eagles' selective response to disturbance to our advantage. One nest in 1982 was used for studying nestling growth as well as parental behavior. If the eagles had responded to us when we went to enter the blind, only fifty yards from the nest tree, as they did when we went to measure the chicks, then we would have recorded a good deal of disturbed rather than natural behavior. Therefore, we tried to fool the eagles. When we boated up to the nest to take measurements of the chicks, we traveled in a direct path at top speed and waved our life jackets over head. When we were going to observe from the blind, we slowly made a wide sweep of the bay and hid everything, including ourselves, under a canvas tarpaulin. The deception worked. Whereas the direct, conspicuous approach always elicited an excited reaction, the sneaky route was tolerated completely. I was able to climb right into my blind, 82 feet up a big spruce, without even flushing the adults. On occasion the parent on the nest and I would look eye to eye, but I was paid no more attention than if I were a squirrel. The situation was, however, unusual, for most pairs on the lake were much more wary. I do not encourage anyone to observe eagles from blinds, as it is generally too risky for the birds. Many a photographer has caused birds to desert their nests because of the disturbing effect of blinds.

Human disturbance is not just a problem in the breeding season; its effect on wintering birds can be serious. Eagles may waste valuable time and energy flying away from people. If the birds are prevented from feeding or gathering food, or if the energy reserves that help them make it through cold periods are used up in escape flights, then some eagles may starve to death.[23] Therefore, some winter refuges have restricted human traffic to minimize disturbance.

Reintroduction Programs

Where eagles have been extirpated, wildlife managers can either choose to do nothing at all and hope that nature restores the species to its former haunts, or they can release birds in an active effort to rebuild the population. Thirteen states and the province of Ontario have chosen the latter option, with New York State, under the direction of Peter Nye, being the leader and innovator in the field. The release technique is a modification of the falconer's practice called hacking. For Bald Eagles, nestlings six to nine weeks old are placed in a man-made nest, usually a cage about seven or eight feet square. They are fed and watched over by hidden caretakers and then released at about twelve weeks old.[24] Food is provided at or near the artificial nest until the young can forage for themselves.

The first Bald Eagle hacking program was initiated in 1976 when two birds were released in New York State. As of 1985, there have been 19 hack locations and an amazing total of 390 Bald Eagles released to the wild. The largest single hack involved 21 eagles. Most of the young birds (82 percent) were obtained as chicks from wild nests: 41 percent from Alaska and 31 percent from four Canadian provinces (Nova Scotia, Ontario, Manitoba, and Saskatchewan). Fifteen percent of the release stock was captive bred, mostly at the Patuxent Wildlife Research Center near Laurel, Maryland.

Hacking towers in New York State (Peter E. Nye).

In Oklahoma, the George Miksch Sutton Avian Research Center has a different approach. Eggs are collected from wild nests in Florida and then hatched in an incubator. The eaglets are raised in captivity until release. The beauty of the program is that if the eggs are collected early enough in the breeding season, the females will lay another clutch. The result is that there is no loss of natural productivity, and the hacking program has a supply of eaglets (28 in the 1985–86 release). This technique, known as double-clutching, is likely feasible only for southern birds, as the laying season of eagles in the northern states and Canada is probably too short to allow re-laying.

Success came early to New York's reintroduction program. In 1980, the two birds released in the first hacking project bred successfully just 84 miles from where they were let go. Most reintroduction programs began in the early eighties, and so it is too early to know if all efforts will be as successful. If one assumes that sexual maturity is at five years of age, then for all programs in 1985 there were 36 individuals that were theoretically available for breeding. As of 1985, seven released eagles (comprising five distinct pairs) have nested. There have been 17 breeding attempts, resulting in 17 fledged offspring.

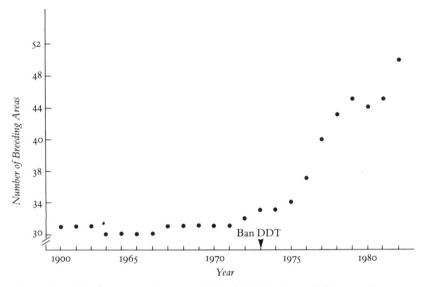

The number of breeding areas (where one pair breeds) in the Greater Yellowstone Ecosystem (northwestern Wyoming, eastern Idaho, and southwestern Montana) increased steadily after the pesticide DDT was banned (data from Swenson et al. 1986).

Contingent largely upon the survival rate of the released birds, it is possible that by 1991 a population may be established that will maintain itself through reproduction and without further input from hacking.

The Future

Many of the problems that eagles have faced historically are still present; however, times have changed. Perhaps more than any nonhuman species in North America, the Bald Eagle has been the subject of legislative action and the focus for environmental reform. That so much interest has been given to a nongame animal is nothing short of remarkable. There has been a tremendous response by public, private, and corporate sectors of society in funding programs to bring back the Bald Eagle. Such efforts have not been unrewarded.

Across the Bald Eagle's range in areas where populations were reduced, the species is beginning to reestablish itself. Since the end of the DDT era in the early seventies, there has been a steady increase in both the number of pairs breeding and the number of young reared per breeding attempt in such areas as Chesapeake Bay, the Greater Yellowstone Ecosystem, and the Great Lakes states. Although a great deal of historical eagle habitat has been made irrevocably unsuitable, populations should continue to grow.

The eagle populations being restored to civilized North America are living in a different environment than the one inhabited by their ancestors

A nestling Bald Eagle, just three days old (Gary R. Bortolotti).

a century or even a few decades ago. Provided there is a clean source of food and a place free of persecution to live their lives and rear young, there is no reason to suspect that the species cannot adapt to the North America of today.

Someday soon we may see eagles return to many of the lakes and rivers where they used to nest. Given current population trends, there is at least hope, hope that was not there less than two decades ago. Gone will be the image of the Bald Eagle as a wilderness monarch; in its place will be a bird that everyone can admire, respect, and enjoy.

Appendix 1: Measurements of Male and Female Bald Eagles of Various Ages

Variable*	Age#	Males Average value	Range	Females Average value	Range
Wing chord†	AD	570	541–589	629	592–664
	SA	581	565–600	610	585–639
	OI	602	579–623	632	600–680
	YI	609	555–651	651	620–683
Primary feather	AD	407	374–437	452	430–472
No. 8	SA	415	397–433	434	414–449
	OI	426	405–454	455	436–486
	YI	441	406–478	472	450–493
Secondary feather	AD	326	307–356	369	342–394
No. 1	SA	337	321–351	363	338–386
	OI	360	313–390	384	353–399
	YI	368	316–398	402	361–430
Tail	AD	255	236–274	289	247–308
	SA	263	254–278	284	279–295
	OI	284	267–308	313	291–350
	YI	313	266–351	329	300–372
Bill length	AD	50.8	47.8–53.3	57.2	53.8–59.9
	SA	50.9	48.3–52.9	54.4	51.2–60.6
	OI	51.7	48.4–54.3	55.2	53.0–58.8
	YI	50.3	41.7–53.6	54.3	50.4–58.7
Bill depth	AD	32.2	29.6–34.6	36.9	34.5–39.2
	SA	32.1	31.3–32.7	35.9	33.8–37.8
	OI	32.4	30.9–34.3	34.9	32.6–37.5
	YI	32.2	29.6–34.2	35.8	33.0–41.2
Hallux claw length	AD	39.8	37.7–41.8	45.7	41.2–48.6
	SA	39.8	38.0–43.0	43.9	41.3–47.8
	OI	40.4	38.0–42.7	45.0	42.6–47.7
	YI	39.1	32.7–42.6	44.2	41.4–48.9

* Measurements were made on 135 museum specimens of birds from Canada, Alaska, and the northern United States (see Bortolotti 1984b). All measurements are in millimeters.
# AD = adult, SA = subadult (i.e., nearly adult in plumage), OI = old immature (i.e., some white in the head, tail, and body), and YI = young immature (all-brown coloration).
† The unflattened length of the folded wing.
The following museums kindly granted G.B. access to their collections: U.S. National Museum of Natural History (Smithsonian Institution), Royal Ontario Museum, National Museum of Natural Sciences (Canada), Manitoba Museum of Man and Nature, Saskatchewan Museum of Natural History, and the University of Michigan Museum of Zoology.

Appendix 2

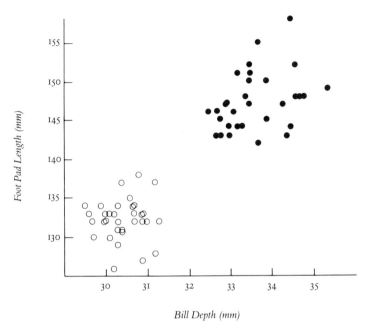

Bill Depth (mm)

The depth of the bill and the length of the foot pad (the span of the fleshy part of the underside of the foot) for male (light circles) and female (dark circles) fledgling Bald Eagles from Besnard Lake. There is no overlap in size between the sexes for either of these measurements (Bortolotti 1984e).

Appendix 3a: Proportion of Prey Species Delivered to Bald Eagle Nests Compared With Proportion of Fish Species Caught in Nets

Species	Delivered to nests[a]		Caught in nets[b]	
	% occurrence	% by weight	% occurrence	% by weight
Cisco (*Coregonus artedii*)	46.7	27.9	35.9	23.4
White sucker (*Catostomus commersoni*)	30.9	39.0	16.6	17.6
Northern pike (*Esox lucius*)	10.0	12.6	8.5	19.2
Burbot (*Lota lota*)	8.6	15.8		
Walleye (*Stizostedion vitreum*)	2.4	3.2	25.5	29.0
Yellow perch (*Perca flavescens*)	0.3	0.1	4.1	0.4
Whitefish (*Coregonus clupeaformis*)	0	0	9.4	10.4
Duck (spp.)	1.0	1.3		

Based on observations of 291 prey items brought to nine nests on Besnard Lake from May to August, 1980 to 1982 (Bortolotti unpublished) and a fisheries survey of Besnard Lake in 1973 (Chen 1974). Note how relatively few walleye (the primary species sought after by human fishermen) are taken by eagles compared with the lake's population.

Appendix 3b: Size of Fish Delivered to Bald Eagle Nests Compared With Size of Fish Caught in Nets

Species	Length (inches)			Weight (grams)			Mean weight of fish caught in gillnet samples (grams)
	Mean	Minimum	Maximum	Mean	Minimum	Maximum	
Cisco	12.0	6	19	272	77	672	384
Sucker	14.7	9	21	574	232	1273	626
Pike	16.8	11	23	574	209	1244	1325
Burbot	17.4	10	31	834	193	2835	(no data)
Walleye	14.8	8	20	602	82	1060	670
Perch	9.5	–	–	190	–	–	57

Data source same as 3a. Note that northern pikes caught by eagles, although not small, are not very large compared with the average size of pike in the lake.

Appendix 4: Number of Adult and Immature Bald Eagles Seen on Boat Censuses of Besnard Lake

Survey		Date	Estimated adult population		Estimated immature population		Total
			No.	%	No.	%	No.
I	1976	May 24–June 1	44.4	64.8	24.1	35.2	68.5
2		May 24–June 1	46.2	63.6	26.4	36.0	72.6
3		June 5–June 8	33.9	68.2	15.8	31.9	49.7
4		June 18–June 25	37.9	59.6	25.7	40.4	63.6
5		June 18–June 25	46.2	69.2	20.6	30.8	66.8
6		July 5–July 10	56.5	55.3	43.6	42.7	102.1
7		July 19–July 23	65.4	54.8	54.0	45.2	119.4
8		August 2–August 5	60.7	62.6	36.2	37.4	96.9
9		August 16–August 20	58.9	53.3	51.7	46.7	110.6
10	1977	May 8–May 13	44.1	68.3	20.4	31.6	64.5
11		May 31–June 7	54.6	76.9	16.4	23.1	71.0
12		June 19–June 24	45.1	59.7	30.5	40.3	75.6
13		July 4–July 9	59.7	54.3	50.2	45.7	109.9
14		July 24–July 27	64.3	57.3	48.0	42.7	112.3
15		July 25–July 29	46.2	46.1	54.0	53.9	100.2
16		August 15–August 18	58.0	76.1	18.2	23.9	76.2
17		August 29–September 1	56.0	77.4	16.3	22.5	72.3
18	1978	May 18–May 21	58.1	83.1	11.8	16.9	69.9
19		June 12–June 14	40.9	69.2	16.2	27.4	59.1
20	1984	May 22–May 25	49.1	69.4	21.6	30.6	70.7
21		June 9–June 11	45.7	74.9	15.3	25.1	61.0
22		July 4–July 11	70.7	77.9	20.1	22.1	90.8
23		August 5–August 7	69.8	67.9	33.0	32.1	102.8

About 44 of the number of adults represent breeding birds (i.e., 22 pairs). Note how the population in July and August increases because of the influx of immatures and nonbreeding adults.

Appendix 5: Clutch Size and Brood size of Bald Eagles at Besnard Lake

Year	Clutch size			Brood Size[a]		
	One egg	Two eggs	Three eggs	One chick	Two chicks	Three chicks
1980	0[b]	6	4	7	11	1
1981	0	7	2	4	11	0
1982	1	12	2	5	13	0
Total	1 (3%)	25 (74%)	8 (24%)	16 (31%)	35 (67%)	1 (2%)
1968–1982	–	–	–	80–86 (43%)	91–98 (53%)	7–8 (4%)

[a] Number of young per brood of banding age (about five weeks old) or older.
[b] Number of nests.

Appendix 6

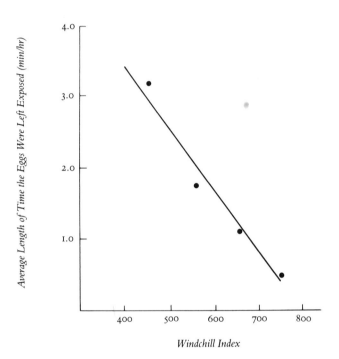

As the windchill index (a combination of temperature and wind speed) increased (the higher the number the colder it feels), the amount of time that incubating eagles in captivity left their eggs exposed decreased (Gerrard et al. 1979).

Appendix 7

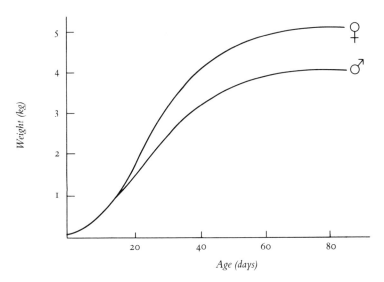

Weight growth curves for male and female nestling Bald Eagles from Besnard Lake. Although the sexes hatch out at the same weight, females are considerably heavier than males at fledging and when mature (Bortolotti 1984d).

Appendix 8

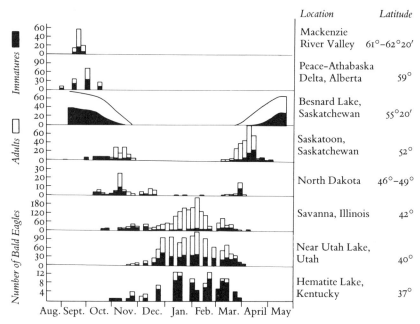

Location	Latitude
Mackenzie River Valley	61°–62°20′
Peace-Athabaska Delta, Alberta	59°
Besnard Lake, Saskatchewan	55°20′
Saskatoon, Saskatchewan	52°
North Dakota	46°–49°
Savanna, Illinois	42°
Near Utah Lake, Utah	40°
Hematite Lake, Kentucky	37°

Observations of Bald Eagles at different latitudes provide a profile of their migration. Immatures are shown by the solid black bars, and adults are shown by the white bars. Immatures tend to go south earlier in fall, go farther south, and return much later in the spring than adults. Eagles at the 61°-62° latitude originate in the Northwest Territories. The Besnard Lake estimates are based on observations over many years. (Data from R. E. Salter personal correspondence; Gerrard unpublished observations; Southern 1964; Edwards 1969; Peterson 1963.)

Appendix 9

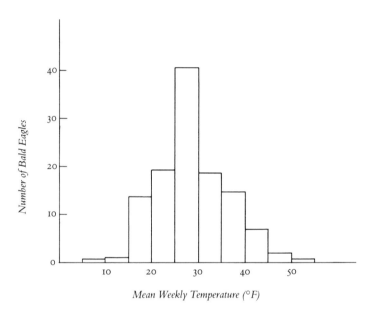

Most Bald Eagles choose wintering areas where the mean daily temperature is just below freezing. Data from Christmas bird counts were used to determine at which mean daily temperature most eagles were found in the western United States. The vertical axis is the number of Bald Eagles per 1000 square miles for the 1970–74 Christmas bird counts.

Appendix 10

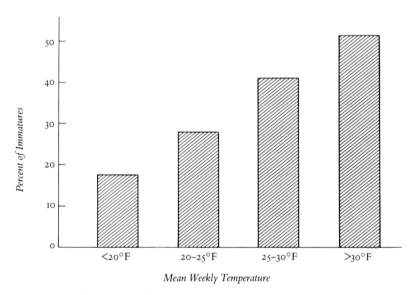

The proportion of wintering Bald Eagles made up of immatures increases at warmer mean daily temperatures. Data from Christmas bird counts in 1974.

Appendix 11: Specifications for Buffer Zones Around Bald Eagle Nests in the Chippewa National Forest, Minnesota*

Each Bald Eagle territory has its own management plan incorporating the following buffer zones.

330-foot zone (100 meters)	– No activity.
660-foot zone (200 meters)	– No activity from February 15 to October 1 and then little.
1320-foot zone (400 meters)	– No activity from February 15 to October 1, no restrictions on activities subsequently.
1320 + -foot zone (400 + meters)	– Includes cases where the 1320- foot zone may be extended up to 2640 feet (800 meters) if justified.

* From Mathisen et al. 1977.

Notes

Chapter 1. Historical Reflections

The excerpt is from Coffin's poem "Bald Eagle," which appeared in *Saturday Review*, May 17, 1952.

1. Abbott 1978; Moseley 1947; Barrows 1912; Roberts 1932; Bailey and Wright 1931; Bailey 1928; Baird et al. 1905 (the Washington Territory referred to includes the present state of Washington, plus parts of the present states of Idaho and Montana); Richardson and Swainson 1831.

2. Grinnell 1920; Musselman 1942.

3. Audubon 1827; Swenson 1983. At the Montana locations mentioned in Swenson's article as lacking eagles in the late 1800s, Bald Eagles are now fairly common in late fall, winter, and early spring, and the birds were probably numerous in the early 1800s.

4. Mowat (1984) describes the incredible numbers of seabirds, shorebirds, and waterfowl that were present along the east coast when Europeans first arrived in North America and the subsequent depletion of those populations.

5. Evermann 1888.

6. Cameron 1905. In one example, near Yellowstone, Cameron notes that "fifty or sixty eagles must have perished annually as a result of wolf control." He is primarily talking about Golden Eagles, but we know that this region is on an important migratory route for Bald Eagles, and they no doubt suffered as well. For a recent account of trap and poison mortality see Bortolotti 1984a.

7. The extent to which Indians hunted eagles elsewhere is uncertain, but it certainly occurred. In South Dakota, on the Rosebud Indian Reservation, Reagan (1908) comments, "It takes a good horse to buy the feathers of one eagle." See also Hearne 1958 for a report of Indians eating eagles.

8. Dawson and Bowles 1909; Dawson 1903. In sections of the states along the east coast, such as in Massachusetts, there had also been a significant decline in eagle numbers (Bagg and Eliot 1937). Even where numbers of breeding eagles remained, as they did near Savannah, Georgia, a report of immatures breeding suggests that the population was stressed by local shooting and egg collecting (Hoxie 1910). Additional documentation of the rarity of the Bald Eagle in major parts of its original range by the early 1900s is available from many other sources, including observations by Francis Harper in a remote section of northern Saskatchewan (Nero 1963). The latter population is a migratory one, and the low numbers found almost certainly reflected pressures on the eagles' wintering grounds.

9. Baynard 1913. There is now no objective evidence to support or refute the contention that there were serious problems with Bald Eagle predation on razorback pigs. In the eighties there are several locations where Bald Eagles are known to feed on pig offal. To what extent Bald Eagles ever preyed on live pigs is uncertain.

10. Taverner 1953.

11. Chatham 1919.

12. It may seem improbable that the slaughter of eagles could have been of such a magnitude to keep the number of birds so low; however, recent population modeling studies by Jim Grier (1980) have emphasized the influence of mortality in determining the size of a population, and they suggest that it is indeed possible for shooting to have kept eagle numbers down.

13. Musselman 1942, 1945, 1949; Swenson 1983.

14. McClelland 1973.

15. It is also possible that a warming trend or the widespread expansion of commercial fishing (which discards rough fish out onto the ice in winter) may have helped the recovery of eagles by increasing successful breeding on the northernmost parts of their range (Swenson 1983; J. M. Gerrard 1983a).

16. Hansen and Hodges 1985.

17. C. L. Broley 1947; M. J. Broley 1952; J. M. Gerrard 1983b. Though habitat destruction was also prominent in that part of Florida, the major reason for the drastic decline in eagle productivity was clearly DDT. Starting in 1947, DDT caused a decrease in eggshell thickness, which coincided with a decrease in Bald Eagle productivity (Anderson and Hickey 1972).

18. Carson 1962.

19. For British Columbia and Alaska see Nickerson 1974; Hodges et al. 1979, 1984; Davies 1985; and for Saskatchewan see Leighton et al. 1979.

20. Abbott 1986; Cline 1986.

Chapter 2. To Know an Eagle

The excerpt is from Carleton's poem "Eagle and Aeroplane," which appeared in *Harper's Weekly*, October 12, 1911.

1. Audubon 1827; Bent 1937.

2. Audubon 1827.

3. Mengel 1953.

4. J. M. Gerrard et al. 1978; Clark 1983; McCollough 1986.

5. W. S. Clark in Virginia observed one wild four-year-old Bald Eagle in adult plumage, and another bird of the same age had a considerable amount of brown flecking in both the head and tail. For an account of plumage coloration see J. M. Gerrard et al. 1978; Clark 1983; and McCollough 1986.

6. Bortolotti and Honeyman 1985. White-tailed Eagles also show consistent patterns of spotting in the tail throughout their lives (Love 1983).

7. Bortolotti 1984b.

8. C. L. Broley 1947; Postupalsky 1976.

9. The term "eagle" is not a scientific one but rather is an arbitrary label for any large bird of prey. For information on eagles of the world see Brown and Amadon 1968 and Brown 1977.

10. See Love 1983.

11. We know so much about the African Fish Eagle because of the devotion of the late Leslie Brown, eagle enthusiast and expert without equal. In addition to his many scientific papers, see his books *The African Fish Eagle*, 1980, and *Eagles of the World*, 1977.

12. See Chapter 4; Knight and Knight 1986; Stalmaster and Gessaman 1984; Hansen 1986.

13. Bortolotti 1984c.

Chapter 3. Flap, Glide, Soar

The line at the beginning of the chapter is from "Lines Written in the Highlands After a Visit to Burns's Country," by John Keats.

1. Brodkorb (1955) weighed the components of an immature female Bald Eagle and found that of her total 4,082 grams, 272 grams were skeleton and 586 grams were contour feathers, excluding the down.

2. Bortolotti 1984b.

3. Imler and Kalmbach 1955; Bortolotti 1984e.

4. Ruppell 1977; Harmata 1984.

5. J. M. Gerrard and Hatch 1983.

6. J. M. Gerrard et al. 1980.

7. Mead 1983.

8. J. M. Gerrard and Gerrard 1982.

Chapter 4. Talons Awaiting

The excerpt is from Thompson's poem "The Captive Eagle," which appeared in *The Cabinet of Natural History and American Rural Sports* 1:239, in 1830, and in *Oölogist* 29:386–387.

1. McEwan and Hirth 1980.

2. Cash et al. 1985; Todd et al. 1982.

3. Gary Bortolotti and Jon Gerrard unpublished data. In a similar vein, Bald Eagles feeding in winter along the Missouri River preferentially catch surface-feeding fish (Steenhof 1976).

4. J. M. Gerrard et al. 1980.

5. Sherrod et al. 1976.

6. Harmata 1984.

7. Knight and Knight 1983.

8. Shea 1973; McClelland et al. 1982.

9. Reimann 1938.

10. F. R. Smith 1936.

11. De Gange and Nelson 1982.

12. Bendire 1892.

13. Campbell 1969.

14. Retfalvi 1965.

15. J. M. Gerrard and Hatch 1983; Harmata et al. 1985.

16. L. B. Dalton, p. 95 in Spencer 1976.

17. Edwards 1969.

18. Kastner 1977.

19. Dekker 1984.

20. Bent 1937.
21. McIlhenny 1932.

Chapter 5. To Find an Eagle and Its Nest

The excerpt is from Percival's poem "To the Eagle," published in 1843 in *The Dream of a Day, and Other Poems* (New Haven: S. Babcock). See also Robinson 1883.

1. Andrew and Mosher 1982; Hatch 1985. Charles Broley (1950) notes, "Once I saw carpenters building a house not ten feet away from a tree in which there were two young eagles, three weeks old. The parents continued to feed and raise the young ones and returned last year again to the same tree! The householder is very proud of his eagles and no one dares molest them."
2. Leighton et al. 1979.
3. Fraser 1981; McEwan and Hirth 1979.
4. Swenson 1975.
5. J. M. Gerrard et al. 1975; Anna Leighton and Douglas W. A. Whitfield unpublished data.
6. East 1944.
7. Howell 1935, 1941, 1954, 1973; C. L. Broley 1950, 1951; J. M. Gerrard 1983a; Noell 1948; M. J. Broley 1952.
8. Abbott 1978.
9. Austin-Smith et al. 1981; Stocek and Pearce 1981; Green 1982.
10. Gainer 1932.
11. Bailey and Wright 1931.
12. Aycock 1972; J. C. Smith 1974.
13. Dubuc and Payne 1977.
14. McIlhenny 1932.
15. Ohmart and Sell 1980; Teryl Grubb personal communication.
16. Swenson 1975; Swenson et al. 1986.
17. Sprunt et al. 1973; Green 1982; David Best of the U.S. Fish and Wildlife Service, and Jim Hammill and Red Evans of the Michigan Department of Natural Resources, in comments to Tom Gause; Sergej Postupalsky and Chuck Sindelar personal communication.
18. Mathisen 1980 and personal communication; Mathisen et al. 1977.
19. Grier 1977, 1982, 1985; McKeating 1985.
20. J. M. Gerrard et al. 1983, and unpublished data.
21. Grubb et al. 1975; Retfalvi 1965; Davies 1985. Davies and colleagues estimate a coastal population of 9,078 adults in British Columbia. The percentage of nonbreeding adults is uncertain, but a reasonable guess is that the estimate likely represents about 4,000 breeding pairs.
22. Nickerson 1974; Hodges and Robards 1982; Hodges et al. 1979.
23. Early 1982; Sherrod et al. 1976.

Chapter 6. A Matter of Space

The excerpt is from McLellan's poem "Ruffed Grouse — Partridge," published in 1886 in *Poems of the Rod and Gun* (New York: Henry Thorpe). See also Thurber 1904.

1. See Walt Whitman's poem "The Dalliance of the Eagles" printed at the beginning of Chapter 8. It is an account of Bald Eagles whirling, interpreted as courtship.
2. We have never seen immatures whirl, although they commonly "toe-touch"—the birds roll over and extend their legs, but there is no locking of talons. This seems more like play than aggression. Immatures may not have the aerial dexterity of the adults (because of their long, wide wings or inexperience), or they may not have the motivation to whirl. Love (1983) also never saw immature White-tailed Eagles whirl, although adults do it.
3. Jon Gerrard and Al Harmata unpublished data.
4. Swenson et al. 1986. We have, however, seen some immatures vigorously chased out of the territory.
5. Mattsson 1974; Jon Gerrard and Gary Bortolotti unpublished data.
6. C. L. Broley 1947; Mattsson 1974; Mahaffy 1981; Swenson et al. 1986. It is difficult to properly assess the size of a territory because aggressive territorial interactions are not common events. We need better data before variation in territory size among populations can be safely attributed to differences in the birds rather than the researchers' methods of measurement.
7. Hancock 1970.
8. Sherrod et al. 1976.
9. A nest can be used by pairs other than the one that built it (probably in most cases after the original occupants have died or abandoned it), and so it is not always safe to assume that the same birds are at the same nest each year.
10. J. M. Gerrard et al. 1980.
11. Newton 1979.
12. J. M. Gerrard et al. 1978.
13. J. M. Gerrard et al. submitted manuscript.
14. Hansen and Hodges 1985.
15. Peter Nye personal communication. It may be difficult to prove conclusively that the three-year-old (a male) did not come in late in the nesting season (Don Hammer, Tennessee Valley Authority).
16. Swenson et al. 1986.
17. Herrick 1932; C. L. Broley 1947.
18. Leighton et al. 1979.
19. Besnard, for example, is a highly productive lake for fish. It comprises two physically distinct sections of water separated by a small narrows. The eastern end of the lake is situated on the Canadian (Precambrian) Shield; the water is deep and blue, and the surrounding terrain is hilly with bare rock outcrops. In contrast, the west end is off the shield; its waters are shallow and murky, and the terrain is flatter and less rocky than in the east. Given its qualities, the west end should have the more productive fish populations, and this has been supported by a fisheries survey (Chen 1974). The density of eagles is 1.5 times higher in the west than in the east, as measured by the number of nests per length of shoreline or nests per area of open water (J. M. Gerrard et al. 1983). See also Whitfield and Gerrard 1985.
20. J. M. Gerrard et al. 1976.
21. Ogden (1975) found that Bald Eagle territoriality had an adverse effect on Osprey reproduction in Florida.

Chapter 7. Eagle Architecture

The excerpt is from McLellan's poem "Ruffed Grouse — Partridge, " published in 1886 in *Poems of the Rod and Gun* (New York: Henry Thorpe). See also Thurber 1904.

1. Some studies of Bald Eagle nesting habitat are J. M. Gerrard et al. 1975 and Whitfield et al. 1974 (Saskatchewan and Manitoba); Hodges et al. 1984 (British Columbia); Swenson et al. 1986 (Greater Yellowstone Ecosystem); Mathisen 1983 (Minnesota); Andrew and Mosher 1982 (Maryland); Grubb 1980 (Washington); McEwan and Hirth 1979 (Florida); Sherrod et al. 1976 (Amchitka Island, Alaska); Hodges and Robards 1982 (southeast Alaska).
2. Personal communication with Anna Leighton and Douglas W. A. Whitfield for Besnard Lake; Barber et al. 1985 for other areas of Saskatchewan.
3. Grubb 1980; Hodges and Robards 1982; Mathisen 1983.
4. Hodges and Robards 1982.
5. Howell 1937; Swenson et al. 1986; personal observation.
6. Sherrod et al. 1976 (Amchitka Island); Grubb 1980 (Washington); Swenson et al. 1986 (Greater Yellowstone Ecosystem). The high breeding density on Amchitka Island may be why these birds do not have alternate nests; perhaps all usable nest sites are occupied by breeders.
7. Newton 1979.
8. Herrick 1932.
9. C. L. Broley 1947.
10. For theories of why birds use greenery see Newton 1979 for raptors and Wimberger 1984 for all species.
11. In more than 1,200 hours of intensive observation from blinds, G.B. has never seen an adult eagle remove any of the putrid carcasses. J.G. has seen it once.
12. A trapper friend of ours, John Hastings, has seen a black bear sitting in an eagle's nest, as have McKelvey and Smith (1979).
13. Nash et al. 1980.
14. Love (1983) believes that White-tailed Eagles may eat vegetation to cushion their digestive tract against sharp bones. He has even seen captive eagles tear and ingest newspaper.
15. Bortolotti 1985.
16. See the *Guinness Book of World Records*; C. L. Broley 1947.
17. Hodges and Robards 1982.
18. J. M. Gerrard et al. 1983.
19. Ohmart and Sell 1980.
20. Sherrod et al. 1976. Observations on Kodiak Island (Dennis Zwieflhofer personal communication), where 40 percent of eagle nests are on the ground in spite of the presence of both foxes and bears, suggest eagles can nest successfully on the ground even where there are mammalian predators.

Chapter 8. A New Generation

American poet Walt Whitman published "The Dalliance of the Eagles" in his book *Leaves of Grass* in 1881.

1. Apparently Whitman never actually saw eagles whirl. The poem was based on a description given to him by John Burroughs, who saw the spectacle in

the early 1860s at Marlboro on the Hudson River (Blodgett and Bradley 1965).

2. P. N. Gerrard et al. 1979.

3. Cindy's behavior is described in detail in J. M. Gerrard et al. 1980.

4. Sherrod et al. 1976; Fraser et al. 1983. The sex composition of the trios was never determined, nor is there any good explanation for why three adults should co-occupy a nest.

5. J. M. Gerrard et al. 1983.

6. For a review of mate retention in birds see Rowley 1983.

7. Harmata 1984.

8. Bent (1937) lists the following egg dates (when eggs were found in nests, not necessarily when they were laid); number of records is in parentheses: Alaska and arctic America—March 24 to June 24 (62); Maine to Michigan—April 1 to 21 (6); New Jersey to Virginia—February 2 to May 27 (75); Georgia and Florida to Texas—October 30 to February 26 (62); Oregon to Mexico—February 18 to April 1 (40).

9. Howell 1937; C. L. Broley 1947.

10. Bortolotti unpublished data.

11. Leighton et al. 1979.

12. Bortolotti unpublished data (derived from hatching dates). Also observed in Florida by C. L. Broley (1947) and Howell (1954).

13. Bent 1937.

14. Newton 1979.

15. C. L. Broley 1947.

16. J. M. Gerrard et al. 1975. Open water in spring is an important component of the habitat of Bald Eagles nesting elsewhere in Saskatchewan (Barber et al. 1985) and in the Greater Yellowstone Ecosystem (Swenson et al. 1986).

17. Al Harmata personal communication.

18. J. M. Gerrard and Whitfield 1979; J. M. Gerrard et al. 1983.

19. Swenson et al. 1986.

20. Sprunt et al. 1973.

21. Pramstaller 1977; P. N. Gerrard et al. 1979.

22. Welty 1975.

23. Herrick 1932; Sherrod et al. 1976; personal observation.

24. P. N. Gerrard et al. 1979; Love 1983.

25. Various lengths of the incubation period have been reported, but 35 days seems to be the most common (e.g., Maestrelli and Wiemeyer 1975).

26. The major cause of infertility of eggs of Bald Eagles in captivity appears to be lack of copulation (Wiemeyer 1981).

27. See Bortolotti et al. 1985 for a discussion of how research activities may disturb nesting eagles. We did not have any deleterious effect on reproduction during our studies.

28. Anderson and Hickey 1972; Wiemeyer et al. 1972; Grier 1982.

29. The embryo is described in detail in Bortolotti 1984d and preserved in spirits in the Royal Ontario Museum, Toronto, Ontario (specimen no. 141311).

Chapter 9. Growing Up

The excerpt is from McLellan's poem "Ruffed Grouse — Partridge," published in 1886 in *Poems of the Rod and Gun* (New York: Henry Thorpe). See also Thurber 1904.

1. For a detailed account of factors influencing sibling competition in Bald Eagles see Bortolotti 1986a and 1986b.
2. J. M. Gerrard et al. 1983.
3. Ankney (1982) found that 64 percent of the first two eggs laid by Snow Geese were males and 72 percent of the last two eggs were females. Similarly, Ryder (1983) found that 64 percent of the first-hatched eggs of Ring-billed Gulls were males and 61 percent of the second-hatched were females. In both geese and gulls males are larger than females. For all three species the percentages are similar, and it is always the larger sex that hatches first.
4. Bortolotti 1984d, 1986a.
5. C. L. Broley 1947.
6. Bortolotti unpublished data.
7. Bortolotti unpublished data. White-tailed Eagles in Greenland also seem to adjust prey deliveries to the number of chicks in the brood but not to the age or size of their offspring (Wille and Kampp 1983). This does not mean that parental effort in foraging is constant, for food availability may vary.
8. Newton 1979.
9. Wiemeyer 1981.
10. Bortolotti 1984d.
11. Chicks begin to thermoregulate at about fifteen days of age. A detailed account of the physical development of the eaglets can be found in Bortolotti 1984d and 1984e.
12. Stewart 1970; Stalmaster and Gessaman 1984.
13. The noted eagle biologist Valerie Gargett, a self-proclaimed housewife and amateur bird-watcher, had a tame Black Eagle under study in Zimbabwe. Although the species is noted for its vigorous attacks on humans near its nests, Valerie could climb to one nest and take eggs and chicks from underneath the female's breast. The eagle was upset and defensive only if her prey were touched.
14. Grubb 1976.
15. M. J. Broley 1952.
16. Bortolotti 1986a.

Chapter 10. Whither the Wind Blows

The excerpt is from McLellan's poem "Ruffed Grouse — Partridge," published in 1886 in *Poems of the Rod and Gun* (New York: Henry Thorpe). See also Thurber 1904.

1. C. L. Broley 1947.
2. J. M. Gerrard et al. 1978; Nye in press; Cline 1986.
3. P. Gerrard et al. 1974; J. M. Gerrard et al. 1978.
4. Harper and Dunstan 1976.
5. Harmata 1984.
6. B. R. McClelland personal communication.

7. Austin-Smith 1985; McCollough 1986; Sindelar and Evans 1976; Swenson et al. 1986; Biosystems Analysis Inc. 1980; Sherrod et al. 1976.
8. Cummings 1976.
9. Senner 1984.
10. Ingram 1965; Ingram et al. 1982.
11. J. M. Gerrard and Gerrard 1982; J. M. Gerrard and Hatch 1983.
12. J. M. Gerrard and Gerrard 1981.
13. Nijssen et al. 1985; Harmata et al. 1985.

Chapter 11. Anything Edible: Bald Eagles in Winter

The poem appeared in *Wilson's American Ornithology* (Boston: Otis, Broaders, and Co.) in 1840. A considerable amount of general information on wintering eagles can be found in Spencer 1976. New information on the question of whether eagles ever kill lambs was provided in an eyewitness account by McEneaney and Jenkins (1983), who saw an adult Bald Eagle attack and kill an apparently healthy lamb that was separated from its ewe. Additional details of the slaughter of eagles from planes can be found in Laycock 1973a.

1. Wright 1953.
2. Knight and Knight 1983.
3. Harmata 1984.
4. Biosystems Analysis Inc. 1980.
5. Hansen 1986.
6. Terry 1976.
7. Stalmaster and Gessaman 1984.
8. Stalmaster and Gessaman 1982.
9. McCollough 1982; 1986.
10. National Wildlife Federation Bald Eagle winter count; Millsap 1986.
11. J. M. Gerrard 1983a.
12. Waste 1982.
13. Stalmaster and Newman 1979.
14. Edwards 1969.
15. Lish 1973.
16. Griffin and Baskett 1985.
17. Fawks 1983. The percentages of immatures given by Fawks are inconsistent (27 percent on one page, 30 percent on another), presumably because they are based on slightly different regions included in the count. Both percentages are lower than the 36 percent reported by Fawks on the National Wildlife Federation count for the Mississippi Region in 1980.
18. Austin-Smith 1985.
19. MacDonald 1981.

Chapter 12. Stewards or Destroyers?

The excerpt is from White's poem "The Deserted Nation," which appeared in *The New Yorker*, October 8, 1966. See also Tufts 1973.

1. Fretwell 1981.
2. Grier 1980.
3. For information on Bald Eagle mortality see review by Fraser (1985), as well as Redig et al. 1983 and Reichel et al. 1984.

4. Shooting is also the largest single cause of mortality for eagles in reintroduction programs (Nye in press).
5. Much of the information in this chapter was obtained from news releases by the U.S. Fish and Wildlife Service and other agencies.
6. Durham 1981; Bortolotti 1984a.
7. Katherine Durham personal communication.
8. Laycock 1973b.
9. Trapping injuries often do not appear to be severe, and so many birds are released on-site. However, soft tissue damage is often serious, and birds have little chance of survival even after treatment in a rehabilitation center (Durham 1981; Redig et al. 1983).
10. Griffin et al. 1980.
11. Although pure DDT cannot be legally sprayed in the United States, pesticides that have DDT as a component can be. There is strong opposition to the use of these compounds, but progress toward banning them is slow.
12. Malley 1985.
13. Researchers at the Chippewa National Forest, Minnesota, deserve special mention for their efforts as pioneers and leaders in the field of eagle management (e.g., see Mathisen et al. 1977 and Appendix 11).
14. McClelland et al. 1983.
15. Sherrod et al. 1976; Stalmaster 1983; Stalmaster and Gessaman 1984.
16. Ohmart and Sell 1980.
17. Stumpf 1977.
18. See reviews by Call 1979 and Olendorff et al. 1980.
19. McCollough 1986.
20. Grubb (1980), Nash et al. (1980), and Bangs et al. (1982), for example, have shown negative impacts of human activities, whereas Fraser et al. (1985) did not. However, the study areas and methods of assessing human disturbance differed among studies.
21. Reported by Bob Hess in *The Eyas* 9(1), 1986.
22. Bortolotti et al. 1985.
23. Stalmaster (1983) presented an elegant model showing the complex interactions between the energy needs and expenditures of Bald Eagles and how human disturbances can be detrimental to a wintering population.
24. Information on reintroduction programs was supplied by Peter Nye (in press); for Oklahoma see Sherrod et al. 1986.

Bibliography

Abbott, J. M. 1978. Chesapeake Bay Bald Eagles. *Delaware Conservation* 22:3–9.
———. 1986. Bald Eagle breeding success in the Chesapeake Bay Region. Unpublished report.
Anderson, D. W., and J. J. Hickey. 1972. Egg shell changes in certain North American birds. Pages 514–540 in *Proceedings of the 15th International Ornithological Congress*, The Hague, Netherlands.
Andrew, J. M., and J. A. Mosher. 1982. Bald Eagle nest site selection and nesting habitat in Maryland. *Journal of Wildlife Management* 46:382–390.
Ankney, C. D. 1982. Sex ratio and egg sequence in Lesser Snow Geese. *Auk* 99:662–666.
Audubon, J. J. 1827. *The birds of America.* Vol. 1. Published by the author, London.
Austin-Smith, P. 1985. Current status of Bald Eagles in Nova Scotia. Pages 39–43 in J. M. Gerrard and Ingram 1985.
———, D. B. Bands, and D. L. Harris. 1981. Bald Eagle management concerns in Nova Scotia. Pages 221–229 in *Proceedings of Bald Eagle Days*. T. N. Ingram (ed.), Eagle Valley Environmentalists, Apple River, Ill.
Aycock, R. Jr. 1972. Eagle and Osprey progress report. Unpublished report.
Bagg, A. C., and S. A. Eliot, Jr. 1937. *Birds of the Connecticut Valley in Massachusetts.* Hampshire Bookshop, Northhampton, Mass.
Bailey, F. M. 1928. *Birds of New Mexico.* New Mexico Department of Game and Fish.
Bailey, A. M., and E. G. Wright. 1931. Birds of Southern Louisiana. *Wilson Bulletin* 43:190–219.
Baird, S. F., T. M. Brewer, and R. Ridgway. 1905. *A history of North American birds.* Vol. 3, *Land birds.* Little, Brown & Co., Boston.
Bangs, E. E., T. N. Bailey, and V. D. Berns. 1982. Ecology of nesting Bald Eagles on the Kenai National Wildlife Refuge, Alaska. Pages 47–54 in Ladd and Schempf 1982.
Barber, S., H. A. Stelfox, and G. Brewster. 1985. Bald Eagle ecology in relation to potential hydro-electric development on the Churchill River, Saskatchewan. Pages 104–113 in J. M. Gerrard and Ingram 1985.
Barrows, W. B. 1912. *Michigan bird life.* Special Bulletin of the Department of Zoology and Physiology, Michigan Agricultural College, Lansing.
Baynard, O. E. 1913. Breeding birds of Alachua County, Florida. *Auk* 30:240–247.
Bendire, C. E. 1892. *Life histories of North American birds.* Smithsonian Institution, Special Bulletin Number 1, Washington, D.C.

Bent, A. C. 1937. *Life histories of North American birds of prey.* Part 1, U.S. National Museum Bulletin 167, Washington, D.C.

Biosystems Analysis Inc. 1980. *Impacts of a proposed Cooper Creek Dam on Bald Eagles.* Study for Seattle City Light, Office of Environmental Affairs.

Bird, D. M. (ed.). 1983. *Biology and management of Bald Eagles and Ospreys.* Harpell Press, Ste. Anne de Bellevue, Quebec.

Blodgett, H. W., and S. Bradley (eds.). 1965. *Walt Whitman leaves of grass.* Comprehensive reader's edition. New York University Press, New York.

Bortolotti, G. R. 1984a. Trap and poison mortality of Golden and Bald Eagles. *Journal of Wildlife Management* 48:1173–1179.

———. 1984b. Sexual size dimorphism and age-related size variation in Bald Eagles. *Journal of Wildlife Management* 48:72–81.

———. 1984c. Age and sex size variation in Golden Eagles. *Journal of Field Ornithology* 55:54–66.

———. 1984d. Physical development of nestling Bald Eagles with emphasis on the timing of growth events. *Wilson Bulletin* 96:524–542.

———. 1984e. Criteria for determining age and sex of nestling Bald Eagles. *Journal of Field Ornithology* 55:467–481.

———. 1985. Frequency of *Protocalliphora avium* (Diptera: Calliphoridae) infestations on Bald Eagles (*Haliaeetus leucocephalus*). *Canadian Journal of Zoology* 63:165–168.

———. 1986a. Influence of sibling competition on nestling sex ratios of sexually dimorphic birds. *American Naturalist* 127:495–507.

———. 1986b. Evolution of growth rates in eagles: sibling competition vs. energy considerations. *Ecology* 67:182–194.

———, J. M. Gerrard, P. N. Gerrard, and D. W. A. Whitfield. 1985. Minimizing investigator-induced disturbance to nesting Bald Eagles. Pages 85–103 in J. M. Gerrard and Ingram 1985.

Bortolotti, G. R., and V. Honeyman. 1985. Flight feather molt of breeding Saskatchewan Bald Eagles. Pages 166–178 in J. M. Gerrard and Ingram 1985.

Brodkorb, P. 1955. Number of feathers and weights of various systems in a Bald Eagle. *Wilson Bulletin* 67:142–143.

Broley, C. L. 1947. Migration and nesting of Florida Bald Eagles. *Wilson Bulletin* 59:3–20.

———. 1950. The plight of the Florida Bald Eagle. *Audubon* 52:24–26.

———. 1951. The plight of the Florida Bald Eagle worsens. *Audubon* 53:139, 141.

Broley, M. J. 1952. *Eagle man.* Pellegrini and Cudahy, Publishers, New York.

Brown, L. 1977. *Eagles of the world.* Universe Books, New York.

———. 1980. *The African Fish Eagle.* Purnell and Sons, Cape Town, South Africa.

——— and D. Amadon. 1968. *Eagles, hawks and falcons of the world.* McGraw-Hill Book Co., New York.

Call, M. 1979. *Habitat management guides for birds of prey.* Bureau of Land Management Technical Note 338, Denver.

Cameron, E. S. 1905. Nesting of the Golden Eagle in Montana. *Auk* 22:158–167.

Campbell, R. W. 1969. Bald Eagle swimming in ocean with prey. *Auk* 86:561.

Carson, R. L. 1962. *Silent spring.* Houghton Mifflin Co., New York.

Cash, K. J., P. J. Austin-Smith, D. Banks, D. Harris, and P. C. Smith. 1985. Food remains from Bald Eagle nest sites on Cape Breton Island, Nova Scotia. *Journal of Wildlife Management* 49:223–225.

Chatham, J. H. 1919. *The Bald Eagle on the Susquehanna.* Altoona Tribune Co.

Chen, M. Y. 1974. *The fisheries biology of Besnard Lake, 1973.* Saskatchewan Department of Tourism and Renewable Resources, Fisheries Technical Report 74-7, Regina.

Clark, W. S. 1983. The field identification of North American eagles. *American Birds* 37:822–826.

Cline, K. W. 1986. *1986 Chesapeake Bay Bald Eagle banding project report.* National Wildlife Federation, Washington, D. C.

Coffin, R. P. T. 1952. Bald Eagle. *Saturday Review* 35:10.

Cummings, H. 1976. Eagles in Wisconsin. In *Proceedings of the Southern Bald Eagle Conference,* Altamonte Springs, Fla.

Davies, R. G. 1985. A note on the population status of Bald Eagles in British Columbia. Pages 63–67 in J. M. Gerrard and Ingram 1985.

Dawson, W. L. 1903. *The birds of Ohio.* Wheaton Publishing Co., Columbus, Ohio.

—— and J. H. Bowles. 1909. *The birds of Washington.* Vol. 2. Occidental Publishing Co., Seattle, Wash.

De Gange, A. R., and J. W. Nelson. 1982. Bald Eagle predation on nocturnal seabirds. *Journal of Field Ornithology* 53:407–409.

Dekker, D. 1984. Migrations and foraging habits of Bald Eagles in east-central Alberta, 1964–1983. *Blue Jay* 42:199–205.

Dubuc, W., and G. C. Payne. 1977. Observations of southern Bald Eagle nesting in southern Louisiana. 1975–1976 Nesting Survey. Unpublished.

Durham, K. 1981. Injuries to birds of prey caught in leghold traps. *International Journal for the Study of Animal Problems* 26:317–328.

Early, T. J. 1982. Abundance and distribution of breeding raptors in the Aleutian Islands, Alaska. Pages 99–111 in Ladd and Schempf 1982.

East, B. 1944. Eagles I have known. *Natural History* 53:8–15.

Edwards, C. C. 1969. Winter behavior and population dynamics of American eagles in Utah. Ph.D. thesis, Brigham Young University, Provo, Utah.

Evermann, B. W. 1888. Birds of Carroll County, Indiana. *Auk* 5:344–351.

Fawks, E. 1983. *Elton Fawks and Bald Eagles.* Privately printed, East Moline, Ill.

Fraser, J. D. 1981. The breeding biology and status of the Bald Eagle on the Chippewa National Forest. Ph.D. thesis, University of Minnesota, St. Paul.

——. 1985. The impact of human activities on Bald Eagle populations — a review. Pages 68–84 in J. M. Gerrard and Ingram 1985.

——, L. D. Frenzel, J. E. Mathisen, and M. E. Shough. 1983. Three adult Bald Eagles at an active nest. *Raptor Research* 17:29–30.

Fraser, J. D., L. D. Frenzel, and J. E. Mathisen. 1985. The impact of human activities on breeding Bald Eagles in north-central Minnesota. *Journal of Wildlife Management* 49:585–592.

Fretwell, S. 1981. Interview with a Bald Eagle. Pages 1–2 in *The Bird Watch.* Bird Populations Institute, Kansas State University, Manhattan, Kans.

Gainer, A. F. 1932. Nesting of the Bald Eagle. *Wilson Bulletin* 44:3–9.

Gerrard, J. M. 1983a. A review of the current status of Bald Eagles in North America. Pages 5–22 in Bird 1983.

——. 1983b. *Charles Broley: An extraordinary naturalist.* White Horse Plains Publishers, Headingley, Manitoba.

————, P. Gerrard, W. J. Maher, and D. W. A. Whitfield. 1975. Factors influencing nest site selection of Bald Eagles in northern Saskatchewan and Manitoba. *Blue Jay* 33:169–176.

Gerrard, J. M., D. W. A. Whitfield, and W. J. Maher. 1976. Osprey–Bald Eagle relationships in Saskatchewan. *Blue Jay* 34:240–247.

Gerrard, J. M., D. W. A. Whitfield, P. Gerrard, P. N. Gerrard, and W. J. Maher. 1978. Migratory movements and plumage of subadult Saskatchewan Bald Eagles. *Canadian Field-Naturalist* 92:375–382.

Gerrard, J. M., P. N. Gerrard, and D. W. A. Whitfield. 1980. Behavior in a non-breeding Bald Eagle. *Canadian Field-Naturalist* 94:391–397.

Gerrard, J. M., P. N. Gerrard, G. R. Bortolotti, and D. W. A. Whitfield. 1983. A 14-year study of Bald Eagle reproduction on Besnard Lake, Saskatchewan. Pages 160–165 in Bird 1983.

Gerrard, J. M., G. R. Bortolotti, E. Dzus, P. N. Gerrard, D. W. A. Whitfield. A boat census for Bald Eagles in north-central Saskatchewan. Submitted manuscript.

Gerrard, J. M., and P. N. Gerrard. 1981. Some observations of Bald Eagles along the upper Mississippi River near Read's Landing, Minnesota. Pages 15–30 in *Proceedings of Bald Eagle Days*. T. N. Ingram (ed.), Eagle Valley Environmentalists, Apple River, Ill.

————. 1982. Spring migration of Bald Eagles near Saskatoon. *Blue Jay* 40:97–104.

Gerrard, J. M., and D. M. Hatch. 1983. Bald Eagle migration through southern Saskatchewan and Manitoba and North Dakota. *Blue Jay* 41:146–154.

Gerrard, J. M., and T. N. Ingram (eds.). 1985. *The Bald Eagle in Canada*. White Horse Plains Publishers, Headingley, Manitoba.

Gerrard, J. M., and D. W. A. Whitfield. 1979. An analysis of the "crash" in eagle productivity in Saskatchewan in 1975. Pages 42–48 in *Proceedings of a Bald Eagle conference on wintering eagles*. T. N. Ingram (ed.), Eagle Valley Environmentalists, Apple River, Ill., Technical Report BED-79.

Gerrard, P., J. M. Gerrard. D. W. A. Whitfield, and W. J. Maher. 1974. Post-fledging movements of juvenile Bald Eagles. *Blue Jay* 32:218–226.

Gerrard, P. N., S. N. Wiemeyer, and J. M. Gerrard. 1979. Some observations on the behavior of captive Bald Eagles before and during incubation. *Raptor Research* 13:57–64.

Green, N. 1982. 1981 status and distribution of nesting Bald Eagles in the conterminous United States. Pages 89–97 in *Proceedings of Bald Eagle Days*. T. N. Ingram (ed.), Eagle Valley Environmentalists, Apple River, Ill.

Grier, J. W. 1977. Quadrat sampling of a nesting population of Bald Eagles. *Journal of Wildlife Management* 41:438–443.

————. 1980. Modeling approaches to Bald Eagle population dynamics. *Wildlife Society Bulletin* 8:316–322.

————. 1982. Ban of DDT and subsequent recovery of reproduction in Bald Eagles. *Science* 218:1232–1235.

————. 1985. History and procedures of surveying for Bald Eagles in northwestern Ontario. Pages 194–200 in J. M. Gerrard and Ingram 1985.

Griffin, C. R. and T. S. Baskett. 1985. Food availability and winter range sizes of immature and adult Bald Eagles. *Journal of Wildlife Management* 49:592–544.

Griffin, C. R., T. S. Baskett, and R. D. Sparrowe. 1980. Bald Eagles and the management program at Swan Lake National Wildlife Refuge. *Transactions of the North American Wildlife and Natural Resources Conference* 45:252–262.

Grinnell, G. B. 1920. Recollections of Audubon Park. *Auk* 37:372.

Grubb, T. G. 1976. Nesting Bald Eagle attacks researcher. *Auk* 93:842–843.

———. 1980. An evaluation of Bald Eagle nesting in western Washington. Pages 87–103 in Knight et al. 1980.

———, D. A. Manuwal, and C. M. Anderson. 1975. Nest distribution and productivity of Bald Eagles in western Washington. *Murrelet* 56:2–6.

Hancock, D. 1970. *Adventure with eagles*. The Wildlife Conservation Centre, Saanichton, British Columbia.

Hansen, A. J. 1986. Fighting behavior in Bald Eagles: a test of game theory. *Ecology* 67:787–797.

——— and J. I. Hodges, Jr. 1985. High rates of nonbreeding adult Bald Eagles in southeastern Alaska. *Journal of Wildlife Management* 49:454–458.

Harmata, A. R. 1984. Bald Eagles of the San Luis Valley, Colorado: Their winter ecology and spring migration. Ph.D. thesis, Montana State University, Bozeman.

———, J. E. Toepfer, and J. M. Gerrard. 1985. Fall migration of Bald Eagles produced in northern Saskatchewan. *Blue Jay* 43:232–237.

Harper, J. F., and T. C. Dunstan. 1976. Dispersal and migration of fledgling Bald Eagles. Pages 94–100 in *Proceedings of Bald Eagle Days 1976*. Eagle Valley Environmentalists, Apple River, Ill.

Hatch, D. 1985. The status of the Bald Eagle in Manitoba. Pages 52–54 in J. M. Gerrard and Ingram 1985.

Hearne, S. 1958. *A journey to the northern ocean*. R. Glover (ed.), Macmillan & Co., Toronto.

Herrick, F. H. 1932. Daily life of the American eagle: Early phase. *Auk* 49:307–323.

Hodges, J. I., J. G. King, and F. C. Robards. 1979. Resurvey of the Bald Eagle breeding population of southeast Alaska. *Journal of Wildlife Management* 43:219–221.

Hodges, J. I., Jr., J. G. King, and R. Davies. 1984. Bald Eagle breeding population survey of coastal British Columbia. *Journal of Wildlife Management* 48:993–998.

Hodges, J. I., and F. C. Robards. 1982. Observations of 3,850 Bald Eagle nests in Southeast Alaska. Pages 37–66 in Ladd and Schempf 1982.

Howell, J. C. 1935. Bald Eagle incubates horned owl's egg. *Auk* 52:79.

———. 1937. The nesting Bald Eagles of southeastern Florida. *Auk* 54:296–299.

———. 1941. Bald Eagle killed by lightning while incubating its eggs. *Wilson Bulletin* 53:42–43.

———. 1954. A history of some Bald Eagle nest sites in east-central Florida. *Auk* 71:306–309.

———. 1973. The 1971 status of 24 Bald Eagle nest sites in east-central Florida. *Auk* 90:678–680.

Hoxie, W. J. 1910. Notes on the Bald Eagle in Georgia. *Auk* 27:454.

Imler, R. H., and E. R. Kalmbach. 1955. *The Bald Eagle and its economic status*. U.S. Department of the Interior, Fish and Wildlife Service, Circular No. 30.

Ingram, T. N. 1965. Wintering Bald Eagles at Guttenberg, Iowa–Cassville, Wisconsin, 1964–1965. *Iowa Bird Life* 35:66–78.

———, T. Brophy, and D. Sherman. 1982. Raptor migrations over Eagle Valley Nature Preserve. Pages 106–112 in *Proceedings of Bald Eagle Days*. T. N. Ingram (ed.), Eagle Valley Environmentalists, Apple River, Ill.

Kastner, J. 1977. *A world of naturalists*. Alfred A. Knopf, New York.

Knight, R. L., G. T. Allen, M. V. Stalmaster, and C. W. Servheen (eds.). 1980. *Proceedings of the Washington Bald Eagle Symposium*. Seattle, Wash.

Knight, S. K., and R. L. Knight. 1983. Aspects of food finding by wintering Bald Eagles. *Auk* 100:477–484.

———. 1986. Vigilance patterns of Bald Eagles feeding in groups. *Auk* 103:263–272.

Ladd, W. N., and P. F. Schempf (eds.). 1982. *Proceedings of a symposium and workshop on raptor management and biology in Alaska and western Canada*. U.S. Fish and Wildlife Service, Anchorage, Alaska.

Laycock, G. 1973a. *Autumn of the eagle*. Charles Scribner's Sons, New York.

———. 1973b. Saving western eagles from traps and zaps. *Audubon* 75:133.

Leighton, F. A., J. M. Gerrard, P. Gerrard, D. W. A. Whitfield, and W. J. Maher. 1979. An aerial census of Bald Eagles in Saskatchewan. *Journal of Wildlife Management* 43:61–69.

Lish, J. W. 1973. Status and ecology of Bald Eagles and nesting Golden Eagles in Oklahoma. M.Sc. thesis, Oklahoma State University, Stillwater.

Love, J. A. 1983. *The return of the Sea Eagle*. Cambridge University Press, Cambridge.

MacDonald, P. R. N. 1981. The age structure and age-related foraging behaviour of the northern Bald Eagle at a wintering site in Nova Scotia. B.Sc. honors thesis, Acadia University, Wolfville, Nova Scotia.

Maestrelli, J. R., and S. N. Wiemeyer. 1975. Breeding Bald Eagles in captivity. *Wilson Bulletin* 87:45–53.

Mahaffy, M. S. 1981. Territorial behavior of the Bald Eagle on the Chippewa National Forest. M.Sc. thesis, University of Minnesota, St. Paul.

Malley, D. F. 1985. Acid rain and its relationship to fish biology and Bald Eagles. Pages 114–138 in J. M. Gerrard and Ingram 1985.

Mathisen, J. E. 1980. *Bald Eagle–Osprey status report*. U.S. Department of Agriculture Forest Service, Chippewa National Forest, Cass Lake, Minnesota.

———. 1983. Nest site selection by Bald Eagles on the Chippewa National Forest. Pages 95–100 in Bird 1983.

———, D. J. Sorenson, L. D. Frenzel, and T. C. Dunstan. 1977. Management strategy for Bald Eagles. *Transactions of the North American Wildlife and Natural Resources Conference* 42:86–92.

Mattsson, J. P. 1974. Interaction of a breeding pair of Bald Eagles with sub-adults at Sucker Lake, Michigan. M.A. thesis, St. Cloud Lake College, St. Cloud, Minnesota.

McClelland, B. R. 1973. Autumn concentrations of Bald Eagles at Glacier National Park. *Condor* 75:121–123.

———, L. S. Young, D. S. Shea, P. T. McClelland, H. L. Allen, and E. B. Spettigue. 1982. The Bald Eagle concentration in Glacier National Park, Montana: Origin, growth, and variation in numbers. *Living Bird* 19:133–155.

———, 1983. The Bald Eagle concentration in Glacier National Park, Montana. Pages 69–77 in Bird 1983.

McCollough, M. A. 1982. Winter feeding of Bald Eagles in Maine. *Proceedings of Bald Eagle Days*. T. N. Ingram (ed.), Eagle Valley Environmentalists, Apple River, Ill.

———. 1986. The post-fledging ecology and population dynamics of Bald Eagles in Maine. Ph.D. thesis, University of Maine at Orono.

McEneaney, T. P., and M. A. Jenkins. 1983. Bald Eagle predation on domestic sheep. *Wilson Bulletin* 95:694–695.

McEwan, L. C., and D. H. Hirth. 1979. Southern Bald Eagle productivity and nest site selection. *Journal of Wildlife Management* 43:585–594.

———. 1980. Food habits of the Bald Eagle in north-central Florida. *Condor* 82:229–231.

McIlhenny, E. A. 1932. The Blue Goose in its winter home. *Auk* 49:279–306.

McKeating, G. 1985. Charles Broley: Eagles then and now in southern Ontario. Pages 25–34 in J. M. Gerrard and Ingram 1985.

McKelvey, R. W., and D. W. Smith. 1979. A black bear in a Bald Eagle nest. *Murrelet* 60:106–107.

Mead, C. 1983. *Bird migration*. Facts on File Publications, New York.

Mengel, R. M. 1953. On the name of the northern Bald Eagle and the identity of Audubon's gigantic "Bird of Washington." *Auk* 65:145–151.

Millsap, B. A. 1986. Status of wintering Bald Eagles in the conterminous 48 states. *Wildlife Society Bulletin* 14:433–440.

Moseley, E. L. 1947. Variations in bird populations of the north-central states due to climate and other changes. *Auk* 64:15–35.

Mowat, F. 1984. *Sea of slaughter*. McClelland and Stewart, Toronto.

Musselman, T. E. 1942. Eagles of western Illinois. *Auk* 59:105–107.

———. 1945. Bald Eagles and Woodcocks in central-western Illinois. *Auk* 62:458–459.

———. 1949. Concentrations of Bald Eagles on the Mississippi River at Hamilton, Illinois. *Auk* 66:83.

Nash, C., M. Pruett-Jones, and G. T. Allen. 1980. The San Juan Islands Bald Eagle nesting survey. Pages 105–115 in Knight et al. 1980.

Nero, R. W. 1963. Bald Eagles of Lake Athabasca, Saskatchewan. *Canadian Audubon* 25:12–14.

Newton, I. 1979. *Population ecology of raptors*. Buteo Books, Vermillion, S. Dak.

Nickerson, P. R. 1974. *The national Bald Eagle survey, 1973 and 1974*. U.S. Department of the Interior, Fish and Wildlife Service, Division of Technical Assistance.

Nijssen, A. L., A. R. Harmata, and J. M. Gerrard. 1985. The initial southward migration of an adult female Bald Eagle. Pages 186–190 in J. M. Gerrard and Ingram 1985.

Noell, G. W. 1948. Bald Eagle captures tern. *Wilson Bulletin* 60:53.

Nye, P. E. A review of Bald Eagle hacking projects and early results in North America. Presented at International Symposium on Raptor Reintroduction, Raptor Research Foundation Annual Meeting, Sacramento, California, November 1985. Proceedings in press.

Ogden, J. C. 1975. Effects of Bald Eagle territoriality on nesting Ospreys. *Wilson Bulletin* 87:496–505.

Ohmart, R. D., and R. J. Sell. 1980. *The Bald Eagle of the Southwest with special emphasis on the breeding population of Arizona.* U.S. Department of the Interior, Water and Power Resources Service.

Olendorff, R. R., R. S. Motroni, and M. W. Call. *Raptor management—the state of the art in 1980.* Bureau of Land Management Technical Note 345, Denver.

Postupalsky, S. 1976. Banded northern Bald Eagles in Florida and other southern states. *Auk* 93:835–836.

Pramstaller, M. E. 1977. Nocturnal, preroosting, and postroosting behavior of breeding adult and young of the year Bald Eagles (*Haliaeetus leucocephalus alascanus*) on the Chippewa National Forest, Minnesota. M.Sc. thesis, University of Minnesota, St. Paul.

Reagan, A. B. 1908. The birds of the Rosebud Indian Reservation, South Dakota. *Auk* 25:462–467.

Redig, P. T., G. E. Duke, and P. Swanson. 1983. The rehabilitation and release of Bald and Golden eagles: A review of 245 cases. Pages 137–147 in Bird 1983.

Reichel, W. L., S. K. Schmeling, E. Cromartie, T. E. Kaiser, A. J. Krynitsky, T. G. Lamont, B. M. Mulhern, R. M. Prouty, C. J. Stafford, and D. M. Swineford. 1984. Pesticide, PCB, and lead residues in necropsy data for Bald Eagles from 32 states—1978–81. *Environmental Monitoring and Assessment* 4:395–403.

Reimann, E. 1938. Bald Eagle takes live fish. *Auk* 55:524–525.

Retfalvi, L. I. 1965. Breeding behavior and feeding habits of the Bald Eagle (*Haliaeetus leucocephalus* L.) on San Juan Island, Washington. M.Sc.F. thesis, University of British Columbia, Vancouver.

Richardson, J., and W. Swainson. 1831. *Fauna Boreali–Americana. Vol. II, The birds.* John Murray, London.

Roberts, T. S. 1932. *The birds of Minnesota.* University of Minnesota Press, Minneapolis.

Robinson, P. 1883. The American eagle in the poets. *Lippincott's Magazine* 6:189–194.

Rowley, I. 1983. Re-mating in birds. Pages 331–360 in *Mate choice.* P. Bateson (ed.), Cambridge University Press, Cambridge.

Ruppell, G. 1977. *Bird flight.* Van Nostrand Reinhold Co., New York.

Ryder, J. P. 1983. Sex ratio and egg sequence in Ring-billed Gulls. *Auk* 100:726–728.

Senner, S. E. 1984. Why count hawks? A Hawk Mountain perspective. *Hawk Mountain News* 62:42–45.

Shea, D. S. 1973. A management-oriented study of Bald Eagle concentrations in Glacier National park. M.S. thesis, University of Montana, Missoula.

Sherrod, S. K., C. M. White, and F. S. L. Williamson. 1976. Biology of the Bald Eagle on Amchitka Island, Alaska. *Living Bird* 15:143–182.

Sherrod, S. K., M. A. Jenkins, W. D. Harken, J. S. Shackford, and L. E. Sherrod. 1986. *Annual report 1985–86.* George Miksch Sutton Avian Research Center, Bartlesville, Okla.

Sindelar, C., and D. L. Evans. 1976. Banding eagles in Wisconsin. Pages 24–26 in *Proceedings of the Southern Bald Eagle Conference,* Altamonte Springs, Fla.

Smith, F. R. 1936. The food and nesting habits of the Bald Eagle. *Auk* 53:301–305.

Smith, J. C. 1974. *Distribution and number of rare, endangered and peripheral species.* Texas Parks and Wildlife Department, Special Wildlife Investigations, Project Number W-103-R-3.

Spencer, D. A. 1976. *Wintering of the migrant Bald Eagle in the lower 48 states.* National Agricultural Chemicals Association, Washington, D.C.

Sprunt, A., IV, W. B. Robertson, Jr., S. Postupalsky, R. J. Hensel, C. E. Knoder, and F. J. Ligas. 1973. Comparative productivity of six Bald Eagle populations. *Transactions of the North American Wildlife and Natural Resources Conference* 38:96–106.

Stalmaster, M. V. 1983. An energetics simulation model for managing wintering Bald Eagles. *Journal of Wildlife Management* 47:349–359.

——— and J. A. Gessaman. 1982. Food consumption and energy requirements of captive Bald Eagles. *Journal of Wildlife Management* 46:646–654.

———. 1984. Ecological energetics and foraging behavior of overwintering Bald Eagles. *Ecological Monographs* 54:407–428.

Stalmaster, M. V., and J. R. Newman. 1979. Perch-site preferences of wintering Bald Eagles in northwest Washington. *Journal of Wildlife Management* 43:221–224.

Steenhof, K. 1976. The ecology of wintering Bald Eagles in southeastern South Dakota. M.Sc. thesis, University of Missouri, Columbia.

Stewart, P. A. 1970. Weight changes and feeding behavior of a captive-reared Bald Eagle. *Bird-Banding* 41:103–110.

Stocek, R. F., and P. A. Pearce. 1981. Status and breeding success of New Brunswick Bald Eagles. *Canadian Field-Naturalist* 95:428–433.

Stumpf, A. 1977. An experiment with artificial raptor hunting perches. *The Bird Watch.* The Bird Populations Institute, Kansas State University, Manhattan, Kans.

Swenson, J. E. 1975. Ecology of the Bald Eagle and Osprey in Yellowstone National Park. M.Sc. thesis, Montana State University, Bozeman.

———. 1983. Is the northern interior Bald Eagle population in North America increasing? Pages 23–34 in Bird 1983.

———, K. L. Alt, and R. L. Eng. 1986. Ecology of Bald Eagles in the Greater Yellowstone ecosystem. *Wildlife Monographs* 95.

Taverner, P. A. 1953. *Birds of Canada.* Musson Book Co., Toronto.

Terry, M. M. 1976. The ethology of a Bald Eagle population wintering along the Missouri River in South Dakota and Nebraska. M.Sc. thesis, University of South Dakota at Springfield.

Thurber, C. 1904. The Bald Eagle. *Birds and Nature* 16:232–236.

Todd, C. S., L. S. Young, R. B. Owen, Jr., F. J. Gramlich. 1982. Food habits of Bald Eagles in Maine. *Journal of Wildlife Management* 46:636–645.

Tufts, R. W. 1973. *The birds of Nova Scotia.* Second edition. Nova Scotia Museum.

Waste, S. M. 1982. Winter ecology of the Bald Eagles of the Chilkat Valley, Alaska. Pages 68–81 in Ladd and Schempf 1982.

Welty, J. C. 1975. *The life of birds.* Second edition. W. B. Saunders Co., Philadelphia.

Whitfield, D. W. A., and J. M. Gerrard. 1985. Correlation of Bald Eagle density with commercial fish catch. Pages 191–193 in J. M. Gerrard and Ingram 1985.

————, W. J. Maher, and D. W. Davis. 1974. Bald Eagle nesting habitat, density, and reproduction in central Saskatchewan and Manitoba. *Canadian Field-Naturalist* 88:399–407.

Wiemeyer, S. N. 1981. Captive propagation of Bald Eagles at Patuxent Wildlife Research Center and introductions into the wild, 1976–1980. *Raptor Research* 15:68–82.

————, B. M. Mulhern, F. J. Ligas, R. J. Hensel, J. E. Mathisen, F. C. Robards, and S. Postupalsky. 1972. Residues of organochlorine pesticides, polychlorinated biphenyls and mercury in Bald Eagle eggs and changes in shell thickness — 1969 and 1970. *Pesticides Monitoring Journal* 6:50–55.

Wille, F., and K. Kampp. 1983. Food of the White-tailed Eagle *Haliaeetus albicilla* in Greenland. *Holarctic Ecology* 6:81–88.

Wimberger, P. H. 1984. The use of green plant material in bird nests to avoid ectoparasites. *Auk* 101:615–618.

Wright, B. S. 1953. The relation of Bald Eagles to breeding ducks in New Brunswick. *Journal of Wildlife Management* 17:55–62.

Index

Photo Credits

Gary R. Bortolotti: pages ii, xvi, 11, 21, 32, 53, 60, 64, 66, 68, 70, 80, 84, 88, 90, 91, 92, 94, 97, 98, 106, 135, 139, 143

Jon M. Gerrard: pages 40, 85, 105

John E. Swedberg: pages 20, 28, 31, 118, 119

Teryl G. Grubb: pages 42, 69

Mark A. McCollough: pages 12, 13, 34

Frank Wille: page 16

Naomi Gerrard: page 75; *line drawings*, pages 23, 110

Peter E. Nye: page 141

NASA: page 78

Courtesy of Clayton White: page 49

Courtesy of New-York Historical Society: page 10